Sam Holyman

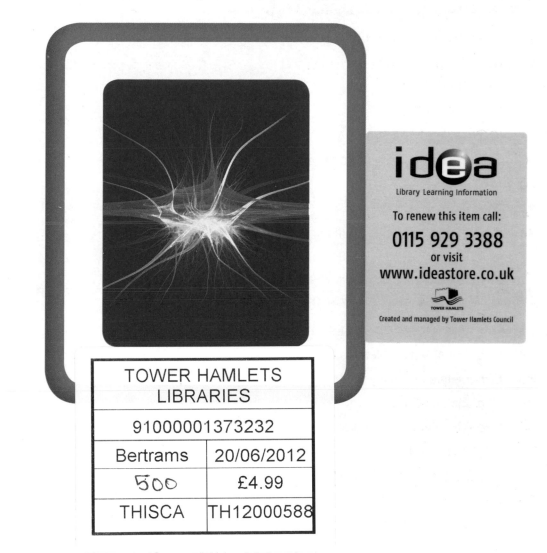

ESSENTIALS

OCR Gateway

GCSE Chemistry
Revision Guide

Contents

Contents

Fundamental Scientific Processes

Scientists carry out **experiments** and collect **evidence** in order to explain how and why things happen. Scientific knowledge and understanding can lead to the **development of new technologies** which have a huge impact on **society** and the **environment**.

Scientific evidence is often based on data that has been collected through **observations** and **measurements**. To allow scientists to reach conclusions, evidence must be **repeatable**, **reproducible** and **valid**.

Models

Models are used to explain scientific ideas and the universe around us. Models can be used to describe:

- a complex idea like how heat moves through a metal
- a system like the Earth's structure.

Models make a system or idea easier to understand by only including the most important parts. They can be used to explain real world observations or to make predictions. But, because models don't contain all the variables, they do sometimes make incorrect predictions.

Models and scientific ideas may change as new observations are made and new data are collected. Data and observations may be collected from a series of experiments. For example, the accepted model of the structure of the atom has been modified as new evidence has been collected from many experiments.

Hypotheses

Scientific explanations are called hypotheses. Hypotheses are used to explain observations. A hypothesis can be tested by planning experiments and collecting data and evidence. For example, if you pull a metal wire you may observe that it stretches. This can be explained by the scientific idea that the atoms in the metal are in layers and can slide over each other. A hypothesis can be modified as new data is collected, and may even be disproved.

Data

Data can be displayed in **tables, pie charts** or **line graphs**. In your exam you may be asked to:

- choose the most appropriate method for displaying data
- identify trends
- use the data mathematically, including using statistical methods, calculating the **mean** and calculating gradients of graphs.

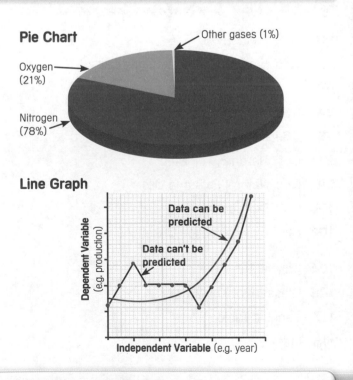

Pie Chart

Other gases (1%)

Oxygen (21%)

Nitrogen (78%)

Line Graph

Data can be predicted

Data can't be predicted

Dependent Variable (e.g. production)

Independent Variable (e.g. year)

Table

Pressure (Atmospheres)	Yield (%) Temperature (°C)			
	250	350	450	550
200	73	50	28	13
400	77	65	45	26

Model • Variable • Data • Hypothesis

Data (Cont.)

Sometimes the same data can lead to different conclusions. For example, data shows that the world's average temperatures have been rising significantly over the last 200 years. Some scientists think this is due to increased combustion of fossil fuels, whilst other scientists think it's a natural change seen before in Earth's history.

Scientific and Technological Development

Every scientific or technological development could have effects that we do not know about. This can give rise to **issues**. An issue is an important question that is in dispute and needs to be settled. Issues could be:

- **Social** – they impact on the human population of a community, city, country, or the world.
- **Environmental** – they impact on the planet, its natural ecosystems and resources.
- **Economic** – money and related factors like employment and the distribution of resources.

- **Cultural** – what is morally right and wrong; a value judgement must be made.

Peer review is a process of self-regulation involving experts in a particular field who **critically examine** the work undertaken. Peer review methods are designed to maintain standards and provide **credibility** for the work that has been carried out. The methods used vary depending on the type of work and also on the overall purpose behind the review process.

Evaluating Information

Conclusions can then be made based on the scientific evidence that has been collected and should try to explain the results and observations.

Evaluations look at the whole investigation. It is important to be able to evaluate information relating to social-scientific issues. When evaluating information:

- make a list of **pluses** (pros)
- make a list of **minuses** (cons)
- consider how each point might **impact on society**.

You also need to consider whether the source of information is reliable and credible and consider opinions, bias and weight of evidence.

Opinions are personal viewpoints. Opinions backed up by valid and reliable evidence carry far more weight than those based on non-scientific ideas. Opinions of experts can also carry more weight

than opinions of non-experts. Information is **biased** if it favours one particular viewpoint without providing a balanced account. Biased information might include incomplete evidence or try to influence how you interpret the evidence.

Fundamental Chemical Concepts

You need to have a good understanding of the concepts (ideas) on these four pages, so make sure you revise this section before each exam.

Atoms

All substances are made up of **atoms**. Atoms have:

- a **positively** charged **nucleus** made of **protons** and **neutrons** (except hydrogen)
- **negatively** charged **electrons** that orbit the nucleus.

Atomic Particle	Relative Charge
Proton	+1
Neutron	0
Electron	-1

An atom contains the same number of electrons (negatively charged particles) and protons (positively charged particles), so each atom is electrically neutral. This means that it has no overall charge.

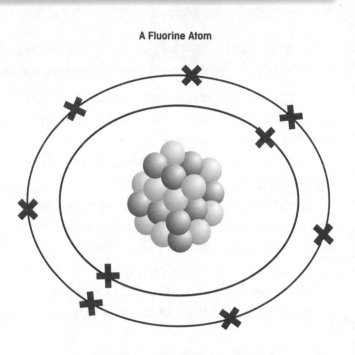

A Fluorine Atom

Key: ● Proton ● Neutron ✖ Electron

Elements and Compounds

An **element** is a substance made up of just one type of atom. Each element is represented by a different chemical symbol, for example:

- Fe represents iron
- Na represents sodium.

These symbols are all arranged in the **periodic table**.

Compounds are substances formed from the atoms of more than one element, which have been joined together by a chemical bond:

- **Covalent** bonds – two atoms **share** a pair of **electrons**. (The atoms in molecules are held together by covalent bonds.)
- **Ionic** bonds – atoms **lose electrons** to become **positive ions** or **gain electrons** to become **negative ions**. The attraction between oppositely charged ions is an **ionic bond**.

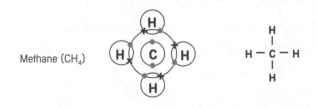

A Covalent Bond between Hydrogen and Carbon in Methane

Methane (CH_4)

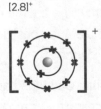

An Ionic Bond between Sodium and Chlorine in Sodium Chloride

Sodium ion, Na^+
$[2.8]^+$

Chloride ion, Cl^-
$[2.8.8]^-$

Key Words Atom • Nucleus • Proton • Neutron • Electron • Element • Compound • Covalent • Ionic

Formulae

Chemical symbols are used with numbers to write **formulae** that represent compounds. Formulae are used to show:
- the different elements in a compound
- the number of atoms of each element in the formula.

If there are brackets around part of the formula, everything inside the brackets is multiplied by the number outside.

Calcium Nitrate

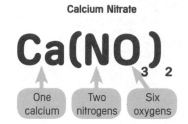

$(NO_3)_2$ means $2 \times NO_3$, i.e. $NO_3 + NO_3$.

Sulfuric Acid

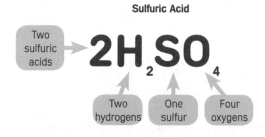

Displayed Formulae

A **displayed formula** shows you the arrangement of atoms in a compound.

A displayed formula shows:
- the different types of atom in the molecule, (e.g. carbon, hydrogen)
- the number of each different type of atom
- the bonds between the atoms.

Ethanol, C_2H_5OH

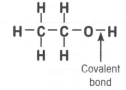

Ethene, C_2H_4

Equations

In a chemical reaction the substances that you start with are called **reactants**. During the reaction, the atoms in the reactants are rearranged to form new substances called **products**.

We use equations to show what has happened during a chemical reaction. The reactants are on the left of the equation and the products are on the right.

No atoms are lost or gained during a chemical reaction so the equation must be balanced: there must always be the same number and type of atoms on both sides of the equation.

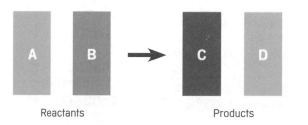

Reactants   Products

Fundamental Chemical Concepts

Writing Balanced Equations

Example 1

1. Write a word equation.

2. Substitute in symbols and formulae.

3. Balance the equation.
   - First, you need to add another MgO to the product side to balance the Os.
   - You now need to add another Mg on the reactant side to balance the Mgs.
   - There are two magnesium atoms and two oxygen atoms on each side – it's balanced.

4. Write a balanced symbol equation.

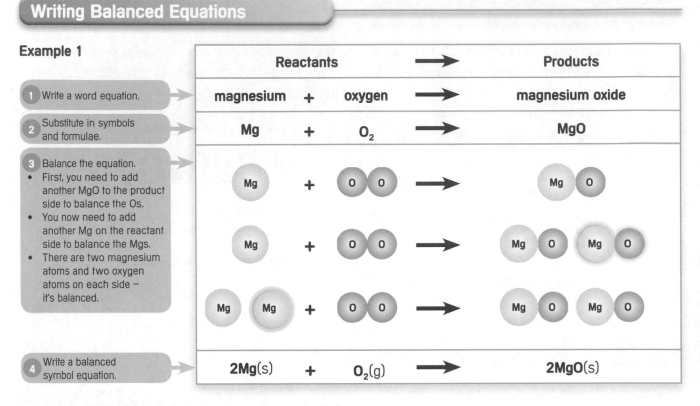

	Reactants		$\longrightarrow$	Products
	magnesium + oxygen		$\longrightarrow$	magnesium oxide
	Mg + O_2		$\longrightarrow$	MgO
	2Mg(s) +	O_2(g)	$\longrightarrow$	2MgO(s)

When you write equations, you may be asked to include the **state symbols**: (aq) for aqueous solutions (dissolved in water), (g) for gases, (l) for liquids and (s) for solids.

(HT) You should be able to balance equations by looking at the formulae (i.e. without drawing the atoms).

1. Write a word equation.

calcium carbonate	+	nitric acid	$\longrightarrow$	calcium nitrate	+	carbon dioxide	+	water

2. Substitute in symbols and formulae.

| $CaCO_3$ | + | HNO_3 | $\longrightarrow$ | $Ca(NO_3)_2$ | + | CO_2 | + | H_2O |

3. Balance the equation.

| $CaCO_3$ | + | $2HNO_3$ | $\longrightarrow$ | $Ca(NO_3)_2$ | + | CO_2 | + | H_2O |

4. Write a balanced symbol equation with state symbols.

| $CaCO_3$(s) | + | $2HNO_3$(aq) | $\longrightarrow$ | $Ca(NO_3)_2$(aq) | + | CO_2(g) | + | H_2O(l) |

Equations can also be written using displayed formulae. These must be balanced too.

| methane | + | oxygen | $\longrightarrow$ | carbon dioxide | + | water |

Fundamental Chemical Concepts

Compounds and their Formulae

Acids

Ethanoic acid	CH_3COOH
Hydrochloric acid	HCl
(HT) Nitric acid	HNO_3
(HT) Sulfuric acid	H_2SO_4

Carbonates

Calcium carbonate	$CaCO_3$
Copper(II) carbonate	$CuCO_3$
(HT) Magnesium carbonate	$MgCO_3$
(HT) Sodium carbonate	Na_2CO_3
(HT) Zinc carbonate	$ZnCO_3$

Chlorides

Ammonium chloride	NH_4Cl
(HT) Calcium chloride	$CaCl_2$
(HT) Magnesium chloride	$MgCl_2$
Potassium chloride	KCl
Silver chloride	$AgCl$
Sodium chloride	$NaCl$

Oxides

Aluminium oxide	Al_2O_3
Copper(II) oxide	CuO
Iron(II) oxide	FeO
(HT) Magnesium oxide	MgO
(HT) Manganese oxide	MnO_2
(HT) Sulfur dioxide	SO_2
(HT) Zinc oxide	ZnO

Hydroxides

Copper(II) hydroxide	$Cu(OH)_2$
Iron(II) hydroxide	$Fe(OH)_2$
(HT) Potassium hydroxide	KOH
(HT) Sodium hydroxide	$NaOH$

Sulfates

(HT) Ammonium sulfate	$(NH_4)_2SO_4$
(HT) Magnesium sulfate	$MgSO_4$
(HT) Potassium sulfate	K_2SO_4
(HT) Sodium sulfate	Na_2SO_4
(HT) Zinc sulfate	$ZnSO_4$

Others

Ammonia	NH_3
Calcium hydrogencarbonate	$Ca(HCO_3)_2$
Carbon dioxide	CO_2
Carbon monoxide	CO
Chlorine	Cl_2
Ethane	C_2H_6
(HT) Ethanol	C_2H_5OH
(HT) Glucose	$C_6H_{12}O_6$
Hydrogen	H_2
Methane	CH_4
Oxygen	O_2
(HT) Silver nitrate	$AgNO_3$
Water	H_2O

Quick Test

1. What are the negative particles in an atom called?
2. Where are the protons and neutrons found in an atom?
3. What is an ion?
4. What three things can displayed formulae tell you?

C1 Making Crude Oil Useful

Fossil Fuels

Crude oil, coal and natural gas are all **fossil fuels**.

Fossil fuels are:

- **formed naturally** over millions of years
- **finite** and **non-renewable** because they are used up much faster than new supplies can be formed – they will be used up in the future
- all easily extracted.

 Crude oil can be used as a source of fuel or chemicals. But it is finite, which means it is being used up much faster than it's being replaced. Scientists are now looking for alternatives for crude oil.

Crude Oil

Crude oil is found in the Earth's crust. It can be pumped to the surface.

Crude oil is transported to refineries through pipelines or in oil tankers.

Accidents can cause oil spills from a tanker and the oil floats on the sea's surface as a **slick**. This can harm wildlife and damage beaches.

The oil can affect lots of wildlife, including birds. The birds' feathers get stuck together and the birds may die.

Detergents are used to break up oil slicks, but these chemicals are toxic and can harm or kill wildlife.

Fractional Distillation

Crude oil is a mixture of many **hydrocarbons**. A hydrocarbon is a molecule that contains **only** **hydrogen** and **carbon** atoms.

Different hydrocarbons have different **boiling points**. This means that crude oil can be separated into useful **fractions** (parts) that contain mixtures of hydrocarbons with similar boiling points. The process used is **fractional distillation**.

The crude oil is heated in a **fractionating column**. The column has a **temperature gradient**, which makes it hotter at the bottom of the column than at the top:

- Fractions with **low boiling points** leave at the **top** of the fractionating column.
- Fractions with **high boiling points** leave at the **bottom** of the fractionating column.

One of the fractions is liquefied petroleum gas (LPG). It contains propane and butane, which are gases at room temperature and are bottled.

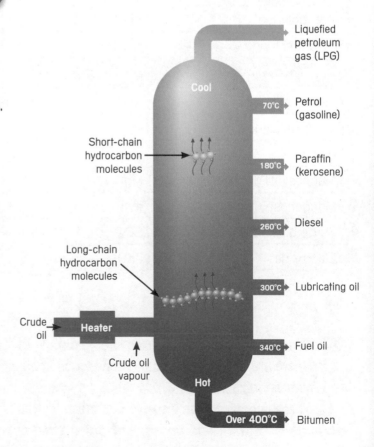

Key Words   Fossil fuel • Non-renewable • Hydrocarbon • Fractional distillation

Cracking

Hydrocarbon molecules can be described as **alkanes** or **alkenes**.

Large alkane molecules can be broken down into smaller, more useful, alkane and alkene molecules.

This industrial process is called **cracking**, and needs a catalyst, a high temperature and a high pressure. In the laboratory, cracking is carried out using the apparatus shown here under atmospheric pressure.

Cracking is used to make more petrol from naphtha. It can also be used to make alkene molecules that may be used to make **polymers**.

There is pressure on these limited resources.

(HT) There isn't enough petrol in crude oil to meet demand. So, cracking is used to change parts of crude oil that can't be used into additional petrol. Crude oil is found in many parts of the world, so oil companies have to work with lots of different countries to extract the oil. Oil is a very valuable resource and is often a source of conflict between nations, and a target for terrorists.

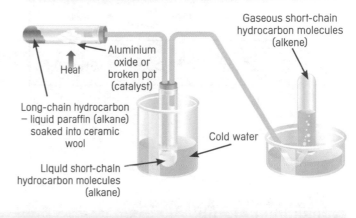

Gaseous short-chain hydrocarbon molecules (alkene)

Aluminium oxide or broken pot (catalyst)

Heat

Long-chain hydrocarbon – liquid paraffin (alkane) soaked into ceramic wool

Cold water

Liquid short-chain hydrocarbon molecules (alkane)

(HT) Forces Between Molecules

In a hydrocarbon molecule there are:

*   strong covalent bonds between the atoms in the molecule
*   weak **intermolecular** forces (forces of attraction between molecules).

The intermolecular forces between longer hydrocarbons are stronger than the forces between shorter hydrocarbons. When a liquid hydrocarbon is boiled, its molecules move faster and faster until all the intermolecular forces are broken and it becomes a gas.

Small molecules have very weak forces of attraction between them and are easy to overcome by heating. It's the differences in their boiling points which allow us to separate a mixture of hydrocarbons (e.g. crude oil) by the process of **fractional distillation**.

Quick Test

1. What does 'non-renewable' mean?
2. What two elements are contained in hydrocarbons?
3. Which fraction contains propane and butane?
4. Long-chain hydrocarbons often undergo further processing before they can be used.
   a) What is cracking?
   (HT) b) Why is cracking carried out?

Key Words　　　　　　Alkane • Alkene • Cracking • Polymer

Combustion

When fuels react with oxygen (in air), they burn and release useful heat energy. This is called **combustion**, and it needs a plentiful supply of oxygen (air).

Many fuels are **hydrocarbons. Complete combustion** of a hydrocarbon, e.g. methane, in air produces carbon dioxide and water.

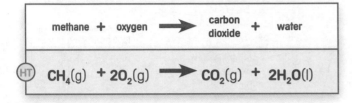

methane + oxygen → carbon dioxide + water

(HT) $CH_4(g) + 2O_2(g) \rightarrow CO_2(g) + 2H_2O(l)$

Complete Combustion

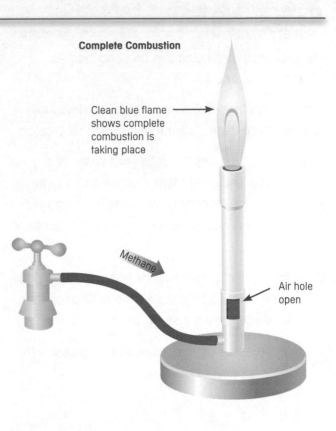

Clean blue flame shows complete combustion is taking place

Methane

Air hole open

Incomplete Combustion

When fuels burn without enough oxygen, then **incomplete combustion** happens. Some heat energy is released, but not as much as complete combustion.

Incomplete combustion of a hydrocarbon produces carbon monoxide (a poisonous gas). This is why gas appliances should be serviced regularly.

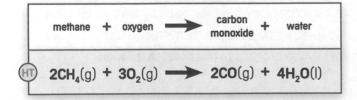

methane + oxygen → carbon monoxide + water

(HT) $2CH_4(g) + 3O_2(g) \rightarrow 2CO(g) + 4H_2O(l)$

When very little oxygen is present, incomplete combustion of a hydrocarbon produces carbon (soot) and water.

methane + oxygen → carbon + water

(HT) $CH_4(g) + O_2(g) \rightarrow C(s) + 2H_2O(l)$

Incomplete Combustion

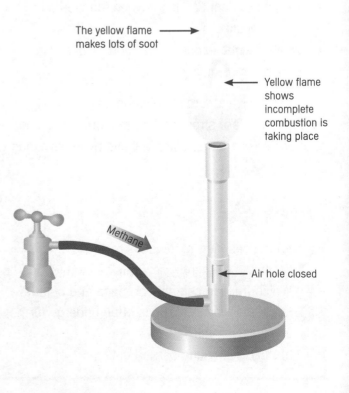

The yellow flame makes lots of soot

Yellow flame shows incomplete combustion is taking place

Methane

Air hole closed

Testing the Products of Combustion

This diagram shows that water and carbon dioxide are made when a hydrocarbon combusts in lots of oxygen.

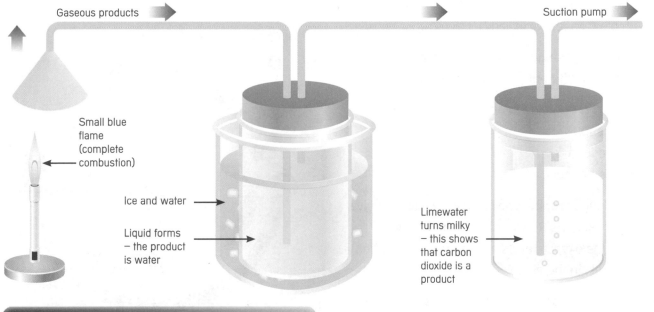

Gaseous products

Suction pump

Small blue flame (complete combustion)

Ice and water

Liquid forms – the product is water

Limewater turns milky – this shows that carbon dioxide is a product

Choosing a Fuel

When choosing a fuel you need to think about:

- **Energy value** – how much energy is released from a gram of fuel?
- **Availability** – how easy is it to get the fuel?
- **Ease of storage** – how easy is it to store the fuel?
- **Cost** – how much fuel do you get for your money?
- **Toxicity** – is the fuel (or its combustion products) poisonous?
- **Pollution** – do the combustion products cause pollution?
- **Ease of use** – is it easy to control and is special equipment needed?

When using a fuel, you want complete combustion to happen because:

- less soot is produced
- more heat energy is released
- no carbon monoxide is made.

HT As the world's population increases and more countries like India and China become industrialised, the demand for **fossil fuels** continues to grow.

Quick Test

1. What type of combustion produces carbon monoxide?
2. What colour is a Bunsen burner flame during complete combustion?
3. Which gas in the air is used for combustion?
4. HT Write a balanced symbol equation for the complete combustion of methane (CH_4).
5. HT Write a balanced symbol equation for the incomplete combustion of methane (CH_4).

C1 Clean Air

The Atmosphere Today

Today, clean dry air contains about:

- 78% **nitrogen**
- 21% **oxygen**
- 1% other gases, including 0.035% carbon dioxide.

The levels of these gases stay about the same. Air also contains different amounts of **water vapour**.

The levels of gas in the **atmosphere** are maintained by:

- **respiration**
- **combustion**
- **photosynthesis**.

All living things **respire**. They take in oxygen and give out carbon dioxide. Respiration and **combustion** decrease the oxygen levels and increase the carbon dioxide levels in the air.

Plants **photosynthesise**: they take in carbon dioxide and release oxygen. Photosynthesis and respiration

balance out, so the levels of carbon dioxide and oxygen in the air stay almost the same.

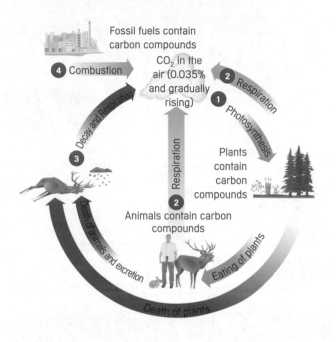

Changing Levels of Gases

The Earth's atmosphere hasn't always been the same as it is today. It has gradually changed over billions of years.

The earliest atmosphere contained ammonia and carbon dioxide. These gases came **from inside the Earth** and were **released** through **volcanoes**.

As plants developed, photosynthesis began and this **reduced** the amount of **carbon dioxide** and **increased** the amount of **oxygen** in the atmosphere.

HT The following is one theory used to explain how Earth's atmosphere evolved:

1. A hot volcanic Earth **released gases** from the crust into the atmosphere. So, the initial atmosphere was made up of ammonia, carbon dioxide and water vapour.

2. As the Earth **cooled**, its surface temperature gradually fell below 100°C and the water vapour **condensed** into liquid water. These newly formed oceans removed some carbon dioxide by dissolving the gas.

3. The levels of nitrogen in the atmosphere increased as **nitrifying bacteria** released nitrogen. This gas is quite unreactive.

4. The level of oxygen in the atmosphere started to increase with the development of primitive plants that could photosynthesise. This **removed carbon dioxide** from the atmosphere, and **added oxygen**.

Air Pollution

The air is becoming increasingly **polluted** with harmful gases due to human actions:

- **Sulfur dioxide** is made when fossil fuels that contain sulfur impurities are burned. It causes acid rain which:
  - kills plants and aquatic life
  - erodes stonework and corrodes ironwork.

- **Carbon monoxide** is a poisonous gas formed from incomplete combustion in a car engine.
- **Oxides of nitrogen** are formed in car engines. They cause photochemical smog and acid rain.

(HT) Nitrogen and oxygen from the air react in the hot car engine to make nitrogen monoxide (NO) and nitrogen dioxide (NO_2).

(HT) Human Influence on the Atmosphere

Three key factors have affected the balance of carbon dioxide that is removed from, and returned to, the atmosphere:

1. **Burning of fossil fuels** is increasing the amount of carbon dioxide in the atmosphere.
2. **Deforestation** on large areas of the Earth's surface means the amount of photosynthesis is reduced so less carbon dioxide is removed from the atmosphere.
3. **Increase in world population** has directly and indirectly contributed to the above factors.

Reducing Pollution

It's important to **reduce** air pollution as much as possible because it can **damage** our surroundings and can **affect** people's health.

Carbon monoxide can be changed into **carbon dioxide** by a **catalytic converter**.

(HT) Catalytic converters contain catalysts which help the polluting chemicals in exhaust gases to react with oxygen. Less harmful gases like carbon dioxide are produced instead. This helps to reduce the amount of pollutants released into the atmosphere.

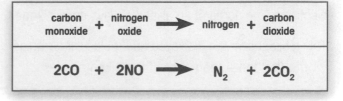

carbon monoxide	+	nitrogen oxide	→	nitrogen	+	carbon dioxide
2CO	+	2NO	→	N_2	+	$2CO_2$

Quick Test

1. What are the two main gases found in dry air?
2. Which two processes decrease oxygen levels in the atmosphere?
3. Which two gases were contained in the Earth's early atmosphere?
4. What types of pollution are caused by oxides of nitrogen?

C1 Making Polymers

Hydrocarbons

Hydrocarbons are compounds that contain only carbon and hydrogen:

- Carbon atoms can make four bonds each.
- Hydrogen atoms can make one bond each.

(HT) To make a hydrocarbon, hydrogen atoms react with carbon atoms to form **covalent** bonds. When this happens, carbon atoms share a pair of electrons with hydrogen atoms to make a covalent bond.

Methane, CH_4

$$
\begin{array}{c}
\quad H \\
\quad | \\
H - C - H \\
\quad | \\
\quad H
\end{array}
$$

Alkanes

When a hydrocarbon chain has only **single covalent bonds**, it is called an **alkane**. All of the carbon atoms make four single covalent bonds. The main chain will contain only single carbon–carbon (C–C) bonds. The name of an alkane always ends in **-ane**.

(HT) Alkanes contain only single covalent bonds – they are described as **saturated** hydrocarbons. (They have the maximum number of hydrogen atoms per carbon atom in the molecule.)

This table shows the displayed and molecular formulae for the first four members of the alkane series.

Alkane	Methane, CH_4	Ethane, C_2H_6	Propane, C_3H_8	Butane, C_4H_{10}
Displayed Formula	$\begin{array}{c} H \\ | \\ H-C-H \\ | \\ H \end{array}$	$\begin{array}{c} H \quad H \\ | \quad | \\ H-C-C-H \\ | \quad | \\ H \quad H \end{array}$	$\begin{array}{c} H \quad H \quad H \\ | \quad | \quad | \\ H-C-C-C-H \\ | \quad | \quad | \\ H \quad H \quad H \end{array}$	$\begin{array}{c} H \quad H \quad H \quad H \\ | \quad | \quad | \quad | \\ H-C-C-C-C-H \\ | \quad | \quad | \quad | \\ H \quad H \quad H \quad H \end{array}$

Alkenes

When a hydrocarbon chain has **one or more double carbon–carbon** (C=C) **covalent bonds**, it's called an **alkene**. Double bonds have two shared pairs of electrons. The name of an alkene always ends in **-ene**.

(HT) Alkenes have at least one double covalent bond, so the carbon atom isn't bonded to the maximum number of hydrogen atoms. Alkenes are described as being **unsaturated**.

This table shows the displayed and molecular formulae for the first three members of the alkene series.

Alkene	Ethene, C_2H_4	Propene, C_3H_6	Butene, C_4H_8
Displayed Formula	$\begin{array}{c} H \qquad H \\ \backslash \quad / \\ C = C \\ / \quad \backslash \\ H \qquad H \end{array}$	$\begin{array}{c} H \qquad\quad H \\ \backslash \qquad | \\ C = C - C - H \\ / \qquad | \\ H \qquad H\ H \end{array}$	$\begin{array}{c} H \qquad\quad H\ \ H \\ \backslash \qquad |\ \ | \\ C = C - C - C - H \\ / \qquad |\ \ | \\ H \qquad H\ H\ H \end{array}$

Covalent • Saturated • Unsaturated

Test for Alkenes

A simple test to distinguish between alkenes and alkanes is to add bromine water:

- Alkenes decolourise bromine water. (The unsaturated alkene reacts with it.)
- Alkanes have no effect on bromine water, i.e. the bromine water stays orange. (The saturated alkane can't react with it.)

HT This reaction is a test for unsaturation. It is an addition reaction between bromine water and the C=C to make a colourless dibromo compound.

Unsaturated Alkene (C=C)   **Saturated Alkane (C−C)**

Bromine water   Bromine water

Polymerisation

The alkenes made by cracking are small molecules which can be used as **monomers**. The double bonds in alkenes are easily broken, so monomers can be joined together to make **polymers** (large, long-chain molecules). Molecules in plastic are called polymers.

When the alkenes join together making a polymer, the reaction is called **polymerisation**. This process needs **high pressure** and a **catalyst**. You could use displayed formulae to show a polymerisation reaction, for example, ethene monomers making poly(ethene):

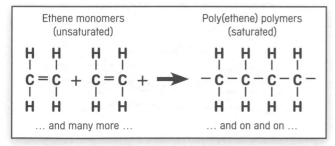

Ethene monomers (unsaturated)   Poly(ethene) polymers (saturated)

… and many more …   … and on and on …

But, it's better to use the standard way of displaying a polymer formula:

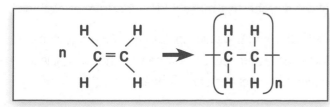

HT Polymerisation involves the reaction of many unsaturated monomer molecules, i.e. alkenes, to form a saturated polymer. You should be able to construct the displayed formula of:

- a **polymer** if you're given the displayed formula of a monomer, e.g. propene monomer to poly(propene) polymer

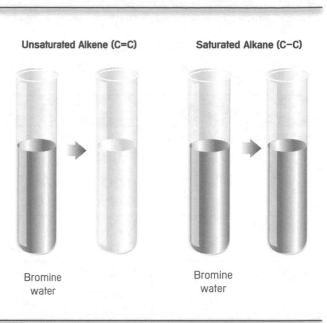

Monomer   **Polymer**

- a **monomer** if you're given the displayed formula of a polymer, e.g. poly(propene) polymer to propene monomer.

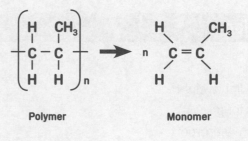

Polymer   **Monomer**

C1 Designer Polymers

Polymers

Different types of **polymers** (plastics) have different properties. These properties relate to the structure of the polymer and means they have many uses. This table lists some polymers' properties and uses.

Polymer	Properties	Uses
Polythene or poly(ethene)	• Light • Flexible • Easily moulded • Can be printed on	• Plastic bags – the plastic is flexible and light. • Moulded containers – the plastic is easily moulded.
Polystyrene	• Light • Poor conductor of heat	• Insulation – the plastic is a poor conductor of heat.
Polyester	• Lightweight • Waterproof • Tough • Can be coloured	• Clothing – the plastic can be made into fibres, is lightweight, tough, waterproof and can be coloured. • Bottles – the plastic is lightweight and waterproof.

(HT) Structure of Plastics

Polymers like PVC are made of tangled very long chain molecules. The atoms are held together by strong **covalent** bonds.

Plastics that have weak forces between polymer molecules (**intermolecular forces**):

- have low melting points
- can be stretched easily because the polymer molecules can slide over each other.

Plastics that have strong forces between the polymer molecules (covalent bonds or cross-linking bridges):

- have high melting points
- are rigid and can't be stretched.

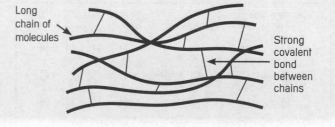

Nylon

Nylon:

- is lightweight
- is tough
- is waterproof
- blocks UV light (harmful sunlight).

Nylon's properties make it ideal for outdoor clothing.

But, although it's waterproof (i.e. keeps you dry) it doesn't let water vapour escape, so it can be uncomfortable if you become hot and start to sweat (perspire).

Gore-Tex®

Gore-Tex® is a breathable material made from nylon. It has all of the advantages of nylon, but it's also treated with a material that allows sweat (water vapour) to escape whilst preventing rain from getting in. This is more comfortable than nylon, as it stops you from getting wet when you sweat.

(HT) In Gore-Tex®, the nylon fibres are coated (laminated) with a membrane of poly(tetrafluoroethene) (PTFE) or polyurethane. This makes the holes in the fabric much smaller.

The coating is used with nylon because it's too weak to be used on its own.

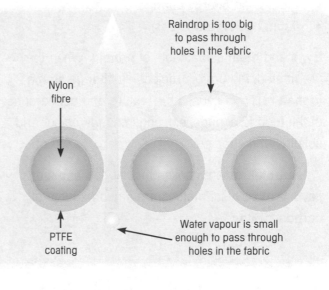

Nylon fibre

PTFE coating

Raindrop is too big to pass through holes in the fabric

Water vapour is small enough to pass through holes in the fabric

Disposing of Polymers

We produce a large amount of plastic waste (polymers) which can be difficult to dispose of and causes litter in the streets.

There are three choices for disposal of plastic waste, but they all have disadvantages. See below.

Research is being carried out on the development of **biodegradable** plastics. These plastics contain special parts which dissolve easily and break up the polymer chain. Biodegradable plastics are already being used in dishwasher detergent tablets.

Using **landfill** sites:	**Burning** polymers:	**Recycling** polymers:
• Most plastics are non-biodegradable, i.e. they will not be broken down by bacteria or decay. • Wastes valuable resources. • Landfill sites get filled up very quickly, i.e. they waste land.	• Produces air pollution. • Some plastics produce toxic fumes when they are burned, for example, burning poly(chloroethene) (PVC) produces hydrogen chloride gas. • Wastes valuable resources.	• Different types of plastic need to be recycled separately – sorting plastics into groups can be time-consuming and expensive.

Quick Test

1. What is a polymer?
2. Name the polymer made from ethene.
3. Give two reasons why polyester is used in clothing.

C1 Cooking and Food Additives

Cooking Food

Cooking food causes a chemical change to take place. When a chemical change happens:

- **new substances** are formed
- there is an **energy change**
- it **can't be reversed** easily.

Eggs and **meat** contain lots of **protein**. When they are heated, the protein molecules **change shape** (**denature**). This causes the texture and appearance of the food to change, e.g. eggs change colour and solidify when heated.

When **potatoes** are cooked they soften and the flavour improves. Potatoes are a good source of **carbohydrates**.

(HT) Potatoes and other vegetables are plants, so their cells have a rigid cell wall.

During cooking, the heat breaks down the cell wall and the cells become soft. Starch grains swell up and are released, so your body can easily digest them.

Eggs and meat contain protein. Denaturing causes the protein molecules to change shape during cooking; it is an irreversible process.

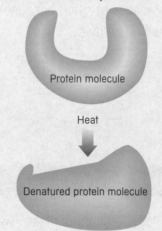

Protein molecule

Heat

Denatured protein molecule

Baking Powder

Baking powder contains **sodium hydrogen carbonate**. When it is heated, it **decomposes** (breaks down) to make sodium carbonate and water, and **carbon dioxide** gas is given off:

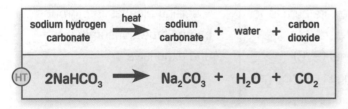

sodium hydrogen carbonate →(heat) sodium carbonate + water + carbon dioxide

(HT) $2NaHCO_3 \longrightarrow Na_2CO_3 + H_2O + CO_2$

Baking powder is added to cake mixture because as the mixture is heated, the carbon dioxide gas that is released causes the cake to **rise**.

Limewater (calcium hydroxide solution) can be used to test for carbon dioxide. If carbon dioxide is present, the limewater will change from colourless to **milky**.

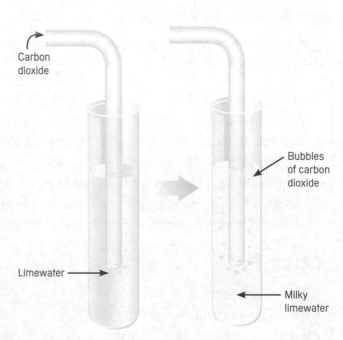

Carbon dioxide

Bubbles of carbon dioxide

Limewater

Milky limewater

Additives

Food **additives** are substances that are put into food to improve it.

A small number of people are **allergic** to certain food additives, i.e. they are harmed by them.

There are different types of food additives:

- **Antioxidants** stop food reacting with oxygen in the air and increase the shelf life.
- **Food colours** improve the appearance of food.
- **Flavour enhancers** bring out the flavour of a food without adding a taste of their own.
- **Emulsifiers** help mix oil and water, which would normally separate, e.g. in mayonnaise.

NUTRITION INFORMATION

Typical Values	Per 100g
Energy	182kJ
Protein	0.8g
Carbohydrates	6.0g
Fat	1.8g
Fibre	0.6g
Sodium	0.3g

Suitable for vegetarians
Suitable for coeliacs

Emulsifiers

Oil and water don't mix, so **emulsifiers** are used. The molecules in an emulsifier have two ends: one end likes to be in water (**hydrophilic**) and the other end likes to be in oil (**hydrophobic**). The emulsifier joins the droplets together and keeps them mixed.

HT The **hydrophilic** end of an emulsifier molecule attracts to the water molecules. The **hydrophobic** end of the emulsifier molecule attracts to the oil molecules. This attraction holds the oil and water molecules together, stopping them from separating.

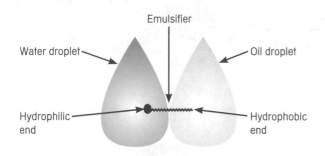

Water droplet — Emulsifier — Oil droplet

Hydrophilic end — Hydrophobic end

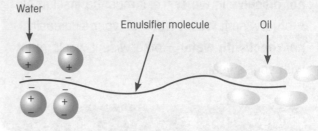

Water — Emulsifier molecule — Oil

Quick Test

1. Give an example of a food that is high in protein.
2. Baking powder is put into cakes to make them rise.
   a) What is the name of the chemical in baking powder?
   b) What chemical change happens when baking powder is heated?
   c) Write a word equation for the reaction when baking powder is heated in a cake.
   d) Which product of this reaction causes cakes to rise?
3. What is meant by 'denaturing'?

C1 Smells

Perfumes

Some cosmetics come from **natural sources**, such as plants and animals. Examples of perfumes from natural sources include lavender, musk and rose.

Cosmetics can also be manufactured. Manufactured perfumes are known as **synthetic perfumes**.

Esters are a family of **compounds** often used as perfumes. An ester is made by reacting an alcohol with an organic acid to produce an ester and water.

Esters can also be used as solvents.

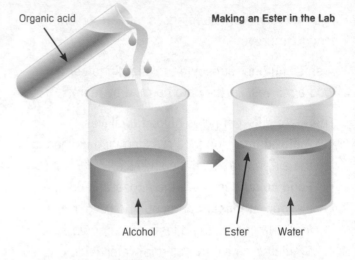

Organic acid

Making an Ester in the Lab

Alcohol   Ester   Water

Properties of Perfume

Smells are made of molecules which travel up your nose and stimulate sense cells.

A perfume must smell nice, and must:
- **evaporate easily** – so it can travel to your nose
- **not be toxic** – so it doesn't poison you
- **not irritate** – otherwise it would be uncomfortable on your skin
- **not dissolve in water** (i.e. it must be **insoluble**) – otherwise it would wash off your skin easily
- **not react with water** – otherwise it would react with your sweat.

(HT) Perfumes are **volatile**, which means they evaporate easily.

The molecules of perfume are held together by weak forces of attraction. The molecules that have lots of energy can easily overcome the weak forces of attraction and escape.

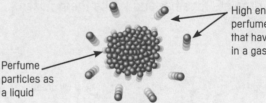

High energy perfume particles that have escaped in a gas state

Perfume particles as a liquid

Testing Perfumes

Perfumes and cosmetics need to be tested to make sure they are safe to use. This testing is sometimes done on animals, although testing on animals has been banned in the EU.

(HT) **Advantage** of animal testing:
- It can prevent humans from being harmed.

Disadvantages of animal testing:
- It's cruel to animals.
- Animals don't have the same body chemistry as humans, so test results might not be useful.

Describing Solutions

Here are some words used to describe substances:

- **Soluble substances** are substances that dissolve in a liquid, e.g. nail varnish is soluble in ethyl ethanoate (nail varnish remover).
- **Insoluble substances** are substances that don't dissolve in a liquid, e.g. nail varnish is insoluble in water.
- A **solvent** is the liquid into which a substance is dissolved, e.g. ethyl ethanoate is a solvent. (An ester can be used as a solvent.)
- The **solute** is the substance that gets dissolved, e.g. the nail varnish is a solute.
- A **solution** is what you get when you mix a solvent and a solute; it will not separate out.

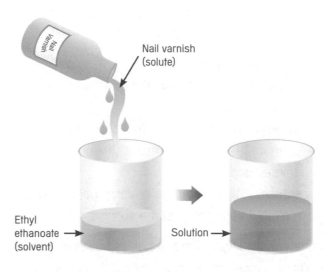

Nail varnish (solute)

Ethyl ethanoate (solvent)

Solution

Solvents

Nail varnish dissolves in nail varnish remover (ethyl ethanoate), but not in water.

HT Nail varnish will not dissolve in water because:
- the attraction between water molecules is stronger than the attraction between water molecules and nail varnish molecules
- the attraction between the molecules in nail varnish is stronger than the attraction between water molecules and nail varnish molecules.

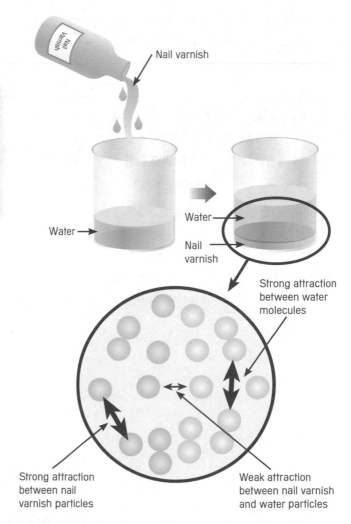

Nail varnish

Water

Water

Nail varnish

Strong attraction between water molecules

Strong attraction between nail varnish particles

Weak attraction between nail varnish and water particles

C1 Paints and Pigments

Paint

Paint is a colloid. Colloids are made of small, solid particles that are mixed well (dispersed but not dissolved) with liquid particles.

Paint is a mixture of:
- **pigment** – a substance that gives paint its colour
- **binding medium** – an oil that sticks the pigment to the surface that it's being painted onto
- **solvent** – thins the thick binding medium and makes it easier to coat the surface.

Paint can be used to **protect** and to **decorate** a surface. It coats the surface with a thin layer and dries when the **solvent evaporates**.

The solvent in **emulsion** paint is water.

In oil-based paints, the pigment is dispersed in an oil (the binding medium). Often, there is a solvent present that dissolves the oil.

HT The particle size of the solids in a colloid must be very small so they stay scattered throughout the mixture. If the particles are too big, they settle down to the bottom of the mixture.

An oil-based paint, such as a gloss paint, dries in two stages:
1. The solvent **evaporates** away.
2. The oil-binding medium reacts with oxygen in the air (an **oxidation reaction**) as it dries to form a hard layer.

Special Pigments

Thermochromic pigments change colour when they are heated or cooled.

These pigments can be used:

- to coat kettles and cups to indicate the temperature
- in mood rings
- in toys and cutlery for babies to warn if food or bath water is too hot.

Phosphorescent pigments glow in the dark. They absorb and store energy and then release it as light when it's dark.

HT A thermochromic pigment can be added to acrylic paints, which makes the paint change through more colours.

The first 'glow in the dark' paints were made using radioactive materials as pigments and were used for things like watches. But, they were dangerous as they exposed people to too much radiation.

Phosphorescent pigments aren't radioactive, so they are much safer to use.

Quick Test

1. Give an example of a cosmetic from a natural source.
2. Why is it important that perfumes do not dissolve in water?
3. What is a colloid?
4. What are the three main parts that make up a paint?
5. What is paint used for?
6. **HT** How does oil-based paint dry?

1 a) Crude oil is a mixture of hydrocarbons.

Explain how fractional distillation separates the fractions in crude oil. **[2]**

....

....

b) What is cracking and why is it used? **[2]**

....

....

c) Explain why small hydrocarbon molecules have a lower boiling point than large hydrocarbon
molecules. **[2]**

....

....

2 Clean air contains a mixture of gases. The proportions of these gases are shown in the pie chart below.

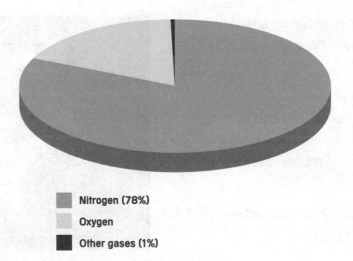

■ Nitrogen (78%)
■ Oxygen
■ Other gases (1%)

a) Which gas makes up most of the air? **[1]**

....

b) What percentage of the air is made up of oxygen? **[1]**

....

c) Give an example of a gas that would be found in 'other gases'. **[1]**

....

3 The Bunsen burner has two flames: the yellow safety flame and the blue heating flame.

a) Explain why a blue Bunsen flame is hotter than a yellow Bunsen flame. **[1]**

....

....

b) Write the word equation for the complete combustion of methane. **[1]**

....

4 Alkenes can be used to make polymers (plastics).

a) What is a monomer? .... **[1]**

b) Describe the conditions needed for the polymerisation of an alkene. **[2]**

....

HT

5 Gore-Tex® is a breathable fabric, used to make outdoor clothes. Explain how a Gore-Tex® coat is waterproof but allows water vapour to escape from inside. **[3]**

....

....

....

6 Emulsifiers are molecules that keep oil and water mixed. Explain how an emulsifier molecule keeps oil and water mixed. **[6]**

✎ The quality of your written communication will be assessed in your answer to this question.

....

....

....

....

....

....

....

....

C2 The Structure of the Earth

Structure of the Earth

The **Earth** is made of a layered structure. It has a:
- thin, rocky **crust**
- **mantle**
- **core** (containing iron).

It's difficult to collect information about the structure of the Earth. The deepest mines and holes drilled into the crust are only a few kilometres into the thick crust.

Scientists have to rely on studying the **seismic waves** (vibrations) caused by **earthquakes** to understand the structure of the Earth.

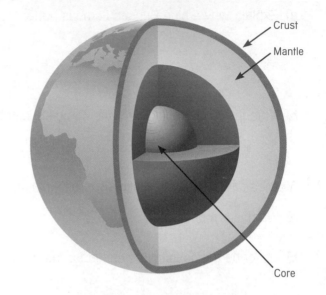

Crust
Mantle
Core

Movement of the Lithosphere

The Earth's **lithosphere** is the relatively cold, rigid outer part of the Earth, made of the crust and top part of the mantle.

The top of the lithosphere is 'cracked' into several large interlocking pieces called **tectonic plates**:
- **Oceanic plates** sit under the ocean.
- **Continental plates** form the continents.

The plates sit on top of the mantle because they are **less dense** than the mantle. Plates move very slowly (about 2.5cm per year). These movements cause **earthquakes** and **volcanoes** at the **boundaries** between plates.

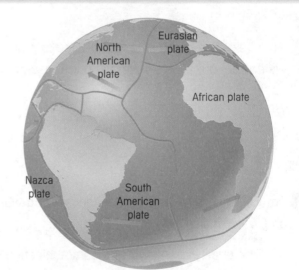

North American plate
Eurasian plate
African plate
Nazca plate
South American plate

Plates Moving Past Each Other

An earthquake will occur along the line where the two plates meet

Key Words   **Crust • Mantle • Core • Seismic wave • Lithosphere • Tectonic plate**

HT What Causes Plates to Move?

Just below the crust, the **mantle** is relatively cold and rigid. At greater depths, the mantle becomes hot and fluid, which means that it can flow. There are **convection currents**, formed by **heat** released from radioactive decay in the core.

Convection currents cause **magma** (molten rock) to **rise** to the surface at the boundaries of plates.

When the molten rock solidifies, new **igneous rock** is formed. This slow movement of the magma causes the plates to **move**.

Oceanic crust has a higher **density** than continental crust. When an oceanic plate collides with a continental plate, it dips down and **slides under** it. This is called subduction. The oceanic plate is **partially re-melted** as it goes under the continental plate.

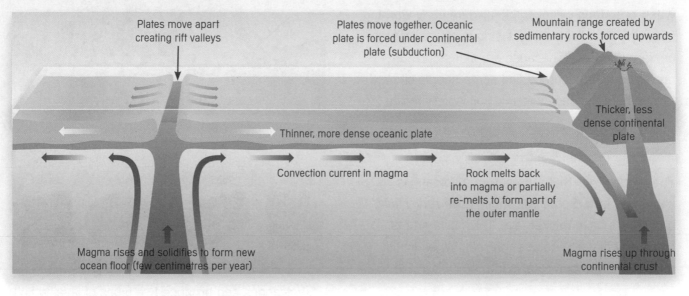

Plates move apart creating rift valleys

Plates move together. Oceanic plate is forced under continental plate (subduction)

Mountain range created by sedimentary rocks forced upwards

Thicker, less dense continental plate

Thinner, more dense oceanic plate

Convection current in magma

Rock melts back into magma or partially re-melts to form part of the outer mantle

Magma rises and solidifies to form new ocean floor (few centimetres per year)

Magma rises up through continental crust

Developing a Theory

Many theories (ideas) have been put forward to explain changes in the Earth's surface. Earth scientists now accept the theory of **plate tectonics**.

HT In 1914, a scientist called Alfred Wegener suggested that the surface of the Earth was changing. He developed the idea that, millions of years ago, all the continents were joined together. Wegener noticed several features on the surface of the Earth:

- The continents look like they would fit together like a **jigsaw**.
- The geology of Scotland and Canada was similar, as was the geology of Africa and South America.
- **Similar animal species** were found on either side of the Atlantic, e.g. caribou in Canada and reindeer in Scandinavia.

This theory is widely accepted as it explains a range of evidence and has been discussed and tested by many scientists.

Initially, Wegener's ideas were not accepted by other scientists. But Wegener's theory was supported by studies in the 1960s which looked at new rock formed at oceanic plate boundaries. The studies showed that:

- the plates are moving apart
- the age of rock increases as you move away from the boundary.

So Wegener's theory of plate tectonics has gradually become accepted.

C2 The Structure of the Earth

Volcanoes

Volcanoes form at places where **magma** (molten rock underneath the Earth's surface) can find its way through weaknesses in the Earth's crust. This is often at plate boundaries or where the crust is very thin. The magma rises through the crust because it has a **lower density** than the crust.

Geologists study volcanoes to help understand the **structure** of the **Earth**. They also aim to **predict** when **eruptions** will occur, to give an early warning to people who live nearby. Living near a volcano can be very dangerous because eruptions can't be predicted with accuracy. But some people choose to live there because volcanic soil is very **fertile**.

HT Geologists are now better able to predict eruptions, but they still can't be 100% accurate.

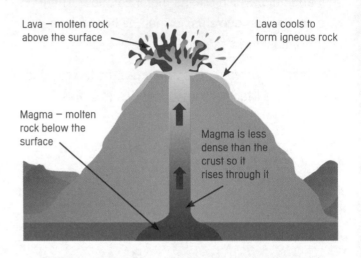

Lava – molten rock above the surface

Lava cools to form igneous rock

Magma – molten rock below the surface

Magma is less dense than the crust so it rises through it

Forming Rock

The molten rock that erupts from a volcano is known as **lava**. Some volcanoes have **runny** lava, and others have **thicker** lava. Thick lava erupts more violently and catastrophically. When liquid rock cools, **igneous rock** is formed.

Igneous rocks are very hard and have interlocking crystals of different sizes:

- **Large crystals** are made when the rock **cools slowly**, as in silica-rich granite and iron-rich gabbro.
- **Small crystals** are made when the rock **cools quickly**, as in silica-rich rhyolite and iron-rich basalt.

Granite   Gabbro   Rhyolite   Basalt

HT Volcanoes can produce two types of lava, which affects the type of eruption:

- **Iron-rich basalt** lava is quite runny and has fairly 'safe' eruptions.
- **Silica-rich rhyolite** is thicker. Thicker lava results in more violent and catastrophic eruptions. Rhyolite lava makes pumice, volcanic ash and bombs.

Quick Test

1. What is the lithosphere made from?
2. Why do tectonic plates float on the mantle?
3. Which two natural disasters happen at plate boundaries?
4. What is igneous rock?
5. Explain how rate of cooling affects crystal size in a piece of igneous rock.
6. HT What are the properties of iron-rich basalt lava?

Materials from Rocks

Many **construction** materials come from rocks found in the Earth's crust:

- Iron and aluminium are extracted from rocks called **ores**.
- Brick is made by baking clay that has been extracted from the Earth.
- Glass, **concrete** and **cement** are all made from sand (small grains of rock).
- Limestone, marble, granite and **aggregates** (gravel) are types of rock extracted from the Earth. These rocks just need to be shaped to be used as building materials.

Limestone is the easiest to shape because it's the softest. Granite is the hardest to shape.

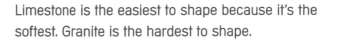 Rocks differ in **hardness** because of the ways in which they were made:

- Limestone is a **sedimentary** rock.
- Marble is a **metamorphic** rock made from limestone that has been put under pressure and heated, which makes it harder.
- Granite is an **igneous** rock.

Limestone, Cement and Concrete

Limestone and marble are mainly made of **calcium carbonate** ($CaCO_3$). When calcium carbonate is heated it breaks down into calcium oxide and carbon dioxide:

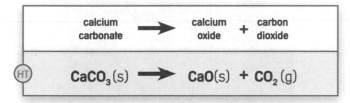

$$\text{calcium carbonate} \longrightarrow \text{calcium oxide} + \text{carbon dioxide}$$

$$CaCO_3(s) \longrightarrow CaO(s) + CO_2(g)$$

This type of reaction is called a **thermal decomposition** reaction; one material breaks down into two or more new substances when heated.

Clay and limestone can be heated together to make cement. Cement can be mixed with sand, gravel (aggregates) and water and allowed to set to make concrete, which is very hard but not very strong. It can be strengthened by allowing it to set around steel rods to reinforce it. **Reinforced concrete** is a **composite** material.

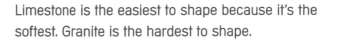 A composite material combines the best properties of each component material. Reinforced concrete combines the **strength** and **flexibility** of the steel bars with the **hardness** of the concrete. Reinforced concrete has many **more uses** than ordinary concrete.

Impact on the Environment

Rock is dug out of the ground in mines and quarries. Mining and quarrying companies have to try to **reduce** their **impact** on the local area and **environment** because mines and quarries can:

- be noisy and dusty
- take up land
- change the shape of the landscape
- increase the local road traffic.

A responsible company will also reconstruct, cover up and restore any area that it has worked on.

C2 Metals and Alloys

Copper

Copper is **extracted** from naturally occurring copper **ore** by **heating** it with carbon:

copper oxide + carbon ⟶ copper + carbon dioxide
$2CuO(s)$ + $C(s)$ ⟶ $2Cu(s)$ + $CO_2(g)$

The process uses lots of energy, which makes it **expensive**. Oxygen is removed from the copper oxide. This process is called **reduction**. It is **cheaper** to **recycle** copper than to extract it from its ore. Recycling also **conserves** the world's limited supply of copper ore and uses less energy.

But recycling copper can be more difficult if it has other metals stuck to it or mixed with it.

If the copper is very **impure**, it can be purified using **electrolysis** (which is an expensive process) before it can be used again.

Electrolysis

Electrolysis uses an **electric current** to break down compounds into simpler substances.

In **electrolysis,** electricity is passed through a liquid or a solution called an **electrolyte**, e.g. copper(II) sulfate solution, to make simpler substances.

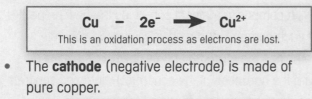

(HT) **Electrodes** are used to allow the electricity to flow through the electrolyte:

*   The **anode** (positive electrode) is made of impure copper.

Cu – $2e^-$ ⟶ Cu^{2+}
This is an oxidation process as electrons are lost.

*   The **cathode** (negative electrode) is made of pure copper.

Cu^{2+} + $2e^-$ ⟶ Cu
This is a reduction process as electrons are gained.

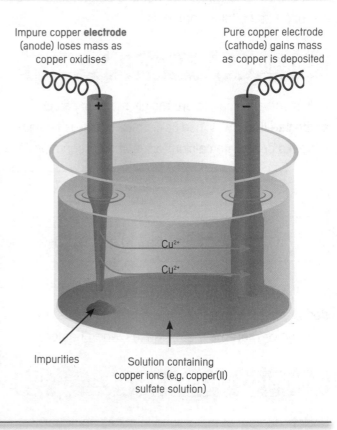

Impure copper **electrode** (anode) loses mass as copper oxidises

Pure copper electrode (cathode) gains mass as copper is deposited

Cu^{2+}

Cu^{2+}

Impurities

Solution containing copper ions (e.g. copper(II) sulfate solution)

Alloys

An **alloy** is a **mixture** of a metal with another element (usually another metal). Bronze and steel are alloys. Alloys improve the **properties** of a metal and make them **more useful** – they are often harder and stronger than the pure metal. For example:

*   **amalgam** (made using mercury) is used for fillings in teeth
*   **solder** (made of lead and tin) is used to join wires
*   **brass** (made of copper and zinc) is used in door handles, coins and musical instruments.

(HT) A **smart alloy** such as nitinol (an alloy of nickel and titanium) can be bent and twisted. Nitinol will return to its original shape when it is heated; it has **shape memory**. This smart alloy is used for the frames of reading glasses.

Recycle • Electrolysis • Electrolyte • Electrode • Alloy • Smart alloy

Materials in a Car

Many different **materials** are used to make cars:

- Nylon **fibre** is used to make the seatbelts because it's strong and flexible.
- **Glass** is used to make the windscreen because it's transparent.
- **Copper** is used for the wiring in the engine because it's a good electrical conductor.
- **Plastic** is used for the trim because it's rigid and doesn't **corrode**.
- **Steel** is used to make the body because it's strong and malleable.
- **Aluminium** is used to make the alloy wheels because it's lightweight and doesn't corrode in moist conditions.

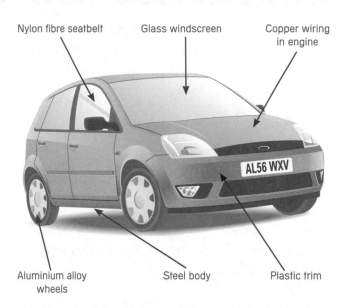

Nylon fibre seatbelt   Glass windscreen   Copper wiring in engine

Aluminium alloy wheels   Steel body   Plastic trim

Properties of Metals and Alloys

The table below compares the properties of aluminium and iron:

Property	Aluminium	Iron
Dense	✗	✔
Magnetic	✗	✔
Resists corrosion	✔	✗
Malleable	✔	✔
Conducts electricity	✔	✔

Aluminium can be mixed with other metals, such as copper and magnesium, to make an **alloy**.

Alloys have different **properties** from the metals that they are made from. These properties make the alloy more **useful**.

For example, **steel** is an alloy of iron and carbon. It is used to make cars because it:

- is harder and stronger than iron
- doesn't corrode as fast as iron.

Aluminium is also used to make car bodies. In comparison to steel, it:

- is lighter
- corrodes less
- is more expensive.

(HT) If aluminium is used to make a car, it will have a longer lifetime because aluminium doesn't corrode.

As aluminium is less dense than steel, the car will be lighter so it will have better fuel economy.

Steel Body   Corrosion   Aluminium Body

C2 Making Cars

Rusting Conditions

Rusting is an example of an **oxidation reaction**, i.e. a reaction where oxygen is added to a substance.

Rusting needs:
- iron
- water
- oxygen (air).

Rusting happens even **faster** when the water is **salty** or is **acid rain**.

Aluminium doesn't react and corrode in air and water. Instead, it quickly forms a **protective layer** of aluminium oxide.

This layer stops any more air or water from coming into contact with the metal. This built-in protection will not flake off.

Oxygen is added to the iron in the presence of water:

iron **+** oxygen **+** water ⟶ **hydrated iron(III) oxide**

Recycling

Most materials used in a car can be **recycled**. Since 2006, the law states that 85% of a car must be able to be recycled; this will increase to 95% in 2015.

Separating all the different materials for recycling can be tricky and time-consuming. But it saves natural resources and avoids disposal problems.

Recycling materials means:
- less quarrying is required
- less energy is used to extract them from ores
- the limited ore reserves will last longer (saves natural resources)
- disposal problems are reduced.

Recycling the plastics and fibres reduces the amount of crude oil needed to make them, and conserves oil reserves.

There are a number of materials in a car that would cause pollution if put into landfill, e.g. lead in the car battery, so recycling also **protects** the **environment**.

Quick Test

1. How are aluminium and iron extracted from the Earth?
2. Give an example of a material that just needs to be shaped before it can be used as a building material.
3. Briefly explain how cement is made.
4. Why is it important to recycle metals?
5. What is an alloy?
6. Why is glass used to make car windscreens?
7. What two elements are needed for iron to rust?

Manufacturing Chemicals: Making Ammonia C2

Ammonia

Ammonia (NH_3) is an alkaline gas made from nitrogen and hydrogen. It can be used to make:

- nitric acid
- **fertilisers** (cheap fertilisers are very important in helping to produce enough food for the growing world population).

The reaction which makes ammonia is a **reversible reaction**. So, nitrogen and hydrogen can form ammonia, and ammonia can decompose to make hydrogen and nitrogen.

Reversible reactions have the symbol ⇌ in their equation to show that the reaction can take place in either direction.

The Haber Process

Ammonia is made on a large scale in the **Haber process**. The reactants are:

- nitrogen (from the air)
- hydrogen (from natural gas or the cracking of crude oil).

Only about 15% of the reactant gases make ammonia. The unreacted gases are recycled. Ammonia is cooled, condensed and then pumped off as a liquid.

nitrogen	+	hydrogen	⇌	ammonia
(HT) N_2(g)	+	$3H_2$(g)	⇌	$2NH_3$(g)

Optimum conditions aren't used as they would be very expensive to maintain, so a compromise is reached:

- The nitrogen and hydrogen mixture is under a **high pressure** of 200 atmospheres.
- The gases are passed over an iron **catalyst** at **450°C**.

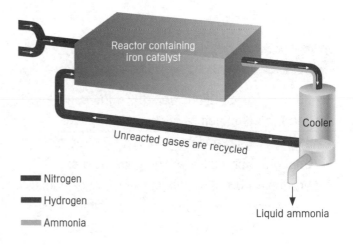

Reactor containing iron catalyst

Cooler

Unreacted gases are recycled

▬ Nitrogen
▬ Hydrogen
▬ Ammonia

Liquid ammonia

Cost and Factors Affecting Cost

The cost of making a new substance depends on:

- the price of energy (gas and electricity)
- labour costs (wages for employees)
- how quickly the new substance can be made (cost of catalyst)
- the cost of starting materials (reactants)
- the cost of equipment needed (plant and machinery).

Factors that affect the cost of making a new substance include:

- the **pressure** – the higher the pressure, the higher the plant cost

- the **temperature** – the higher the temperature, the higher the energy cost
- the **catalysts** – catalysts can be expensive to buy, but production costs are reduced because they increase the rate of reaction
- the **number of people** needed to operate machinery – automation reduces the wages bill
- the **amount of unreacted material** that can be recycled – recycling reduces costs.

C2 Manufacturing Chemicals: Making Ammonia

Interpreting Data

You need to be able to interpret data percentage **yield** in reversible reactions.

Example

The graph and table show how temperature and pressure affect the rate of reaction in the Haber process.

The information given allows you to work out that the yield falls when temperature is increased, and that the yield increases as pressure increases.

(HT) You may also be asked to interpret data on other industrial processes in terms of rate, percentage yield and cost.

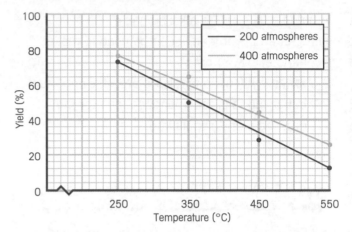

Yield (%) Pressure (Atmospheres)	Temperature (°C)			
	250	350	450	550
200	73	50	28	13
400	77	65	45	26

(HT) Economic Considerations

Economic considerations determine the conditions used in the manufacture of chemicals:

- The **percentage yield** achieved must be high enough to produce enough daily yield of product (a low percentage yield is acceptable providing the reaction can be repeated many times with **recycled starting materials**).
- The **rate of reaction** must be high enough to produce enough daily **yield** of product.
- The **optimum conditions** should be used to give the most economical reaction (this could mean a slower reaction or a lower percentage yield at a lower cost).

Economics of the Haber Process

It's important that the **maximum amount** of ammonia is made in the **shortest possible time** at a **reasonable cost**. This requires a **compromise**.

For the Haber process:

- a low temperature increases yield but the reaction is too slow
- a high pressure increases yield but becomes more expensive as yield increases
- a catalyst increases the rate of reaction but doesn't change the percentage yield.

So, a compromise is reached of:

- temperature of 450°C
- pressure of 200 atmospheres
- catalyst of iron.

This gives a fast reaction with an acceptable percentage yield.

Acids, Bases and Indicators

Indicators are chemicals that change colour to show changes in pH. Some indicators, e.g. litmus, have only two colours; others, e.g. **universal indicator**, have a range of colours over different pH values. **Acids** are substances with a pH of **less than 7**. **Bases** are the oxides and hydroxides of metals, with a pH of **greater than 7**. Acids turn litmus **indicator** red and bases turn litmus indicator blue.

Soluble bases (chemicals with a pH greater than 7 and that dissolve in water) are called **alkalis**.

You can find the pH of a solution by using universal indicator. You can add a few drops of universal indicator to the solution and compare the resulting colour against a **pH colour chart**.

| 0 | 1 | 2 | 3 | 4 | 5 | 6 | 7 | 8 | 9 | 10 | 11 | 12 | 13 | 14 |

Strongly acidic   Weakly acidic   →Neutral← Weakly alkaline   Strongly alkaline

Neutralisation

Metal oxides and metal hydroxides are bases. When they are added to acids in the correct amounts, they can cancel each other out. This is called **neutralisation** because the resulting solution has a **neutral** pH of 7.

acid + base ➡ salt + water

As an **acid** is added to an **alkali**, the **pH** of the solution **decreases** because the acid neutralises the alkali to reach pH 7.

As an **alkali** is added to an **acid**, the **pH** of the solution **increases** because the alkali neutralises the acid to reach pH 7.

Acids can also be neutralised by **carbonates** to produce a **salt**, water and carbon dioxide gas.

acid + carbonate ➡ salt + water + carbon dioxide

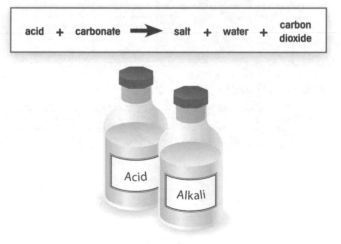

Naming Salts

The **first name** of a salt comes from the name of the **base or carbonate** used, for example:
- **sodium** hydroxide will make a **sodium** salt
- **copper** oxide will make a **copper** salt
- **calcium** carbonate will make a **calcium** salt
- **ammonia** will make an **ammonium** salt.

The **second name** of the salt comes from the **acid** used, for example:
- hydro**chlor**ic acid will produce a **chlor**ide salt
- **sulf**uric acid will produce a **sulf**ate salt
- **nitr**ic acid will produce a **nitr**ate salt
- **phosph**oric acid will produce a **phosph**ate salt.

For example, neutralising **potassium** hydroxide with **nitric** acid will produce **potassium nitrate**.

C2 Acids and Bases

More on Neutralisation

Alkalis in solution contain **hydroxide ions**, **OH⁻**(aq).

Acids in solution contain **hydrogen** ions, **H⁺**(aq).

The pH of a solution is a measure of the concentration of H⁺ ions.

(HT) Neutralisation can be described using the ionic equation:

$$H^+(aq) \ + \ OH^-(aq) \ \longrightarrow \ H_2O(l)$$

(HT) Producing Salts

You should be able to construct any of the following word equations and balanced symbol equations for producing salts.

Acid + Base

	Hydrochloric Acid (HCl)	Sulfuric Acid (H₂SO₄)	Nitric Acid (HNO₃)
Sodium Hydroxide (NaOH)	$NaOH + HCl \rightarrow NaCl + H_2O$	$2NaOH + H_2SO_4 \rightarrow Na_2SO_4 + 2H_2O$	$NaOH + HNO_3 \rightarrow NaNO_3 + H_2O$
Potassium Hydroxide (KOH)	$KOH + HCl \rightarrow KCl + H_2O$	$2KOH + H_2SO_4 \rightarrow K_2SO_4 + 2H_2O$	$KOH + HNO_3 \rightarrow KNO_3 + H_2O$
Copper(II) Oxide (CuO)	$CuO + 2HCl \rightarrow CuCl_2 + H_2O$	$CuO + H_2SO_4 \rightarrow CuSO_4 + H_2O$	$CuO + 2HNO_3 \rightarrow Cu(NO_3)_2 + H_2O$
Ammonia (NH₃)	$NH_3 + HCl \rightarrow NH_4Cl$	$2NH_3 + H_2SO_4 \rightarrow (NH_4)_2SO_4$	$NH_3 + HNO_3 \rightarrow NH_4NO_3$

Acid + Carbonate

	Hydrochloric Acid (HCl)	Sulfuric Acid (H₂SO₄)	Nitric Acid (HNO₃)
Sodium Carbonate (Na₂CO₃)	$Na_2CO_3 + 2HCl \rightarrow 2NaCl + H_2O + CO_2$	$Na_2CO_3 + H_2SO_4 \rightarrow Na_2SO_4 + H_2O + CO_2$	$Na_2CO_3 + 2HNO_3 \rightarrow 2NaNO_3 + H_2O + CO_2$
Calcium Carbonate (CaCO₃)	$CaCO_3 + 2HCl \rightarrow CaCl_2 + H_2O + CO_2$	$CaCO_3 + H_2SO_4 \rightarrow CaSO_4 + H_2O + CO_2$	$CaCO_3 + 2HNO_3 \rightarrow Ca(NO_3)_2 + H_2O + CO_2$

Quick Test

1. Which elements are found in ammonia?
2. Describe the conditions for the Haber process.
3. What is a reversible reaction?
4. What is an acid?
5. What is an alkali?
6. Write down the general word equation for the reaction between a metal carbonate and an acid.

Fertilisers

Fertilisers are chemicals that give plants **essential chemical elements** needed for growth. Fertilisers:
- make crops grow faster and bigger
- increase the crop **yield**.

As world populations rise, fertilisers can increase food supply but can also cause problems such as the death of animals in waterways. This is known as **eutrophication**.

The three main essential elements found in fertilisers are **nitrogen (N)**, **phosphorus (P)** and **potassium (K)**. **Urea** can also be used as a fertiliser.

Fertilisers must be **soluble in water** so that they can be taken in by the roots of plants in solution.

HT Fertilisers **increase crop yield** by:
- **replacing essential elements** in the soil that have been used up by a previous crop
- **providing nitrogen** as soluble nitrates which are used by the plant to make protein for growth.

Making Fertilisers

Some fertilisers can be manufactured by **neutralising** an **acid** with an **alkali**:
- **Ammonium sulfate** – neutralise sulfuric acid with ammonia.
- **Ammonium nitrate** – neutralise nitric acid with ammonia.
- **Ammonium phosphate** – neutralise phosphoric acid with ammonia.
- **Potassium nitrate** – neutralise nitric acid with potassium hydroxide.

You should be able to label the apparatus needed to make a fertiliser by neutralisation:
- Burette
- Measuring cylinder
- Filter funnel

HT A fertiliser, e.g. potassium nitrate, can be made by producing a salt from neutralisation:

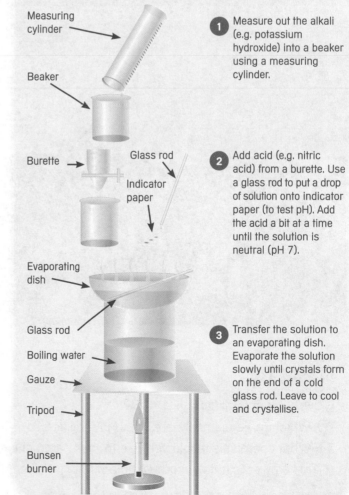

Measuring cylinder

Beaker

Burette

Glass rod

Indicator paper

Evaporating dish

Glass rod

Boiling water

Gauze

Tripod

Bunsen burner

1. Measure out the alkali (e.g. potassium hydroxide) into a beaker using a measuring cylinder.

2. Add acid (e.g. nitric acid) from a burette. Use a glass rod to put a drop of solution onto indicator paper (to test pH). Add the acid a bit at a time until the solution is neutral (pH 7).

3. Transfer the solution to an evaporating dish. Evaporate the solution slowly until crystals form on the end of a cold glass rod. Leave to cool and crystallise.

C2 Fertilisers and Crop Yield

Eutrophication

Eutrophication is when the overuse of fertilisers changes the ecosystem in lakes, rivers and streams.

HT

1. Fertilisers used by farmers may be washed into lakes and rivers (run-off). This increases the levels of nitrates and phosphates in the water and more simple algae grow.

2. The algal bloom blocks off sunlight to other plants, causing them to die and rot.

3. Aerobic bacteria feed on the dead organisms and increase in number. They quickly use up the oxygen until nearly all the oxygen is removed. There isn't enough oxygen left to support the larger organisms, such as fish and other aquatic animals, so they suffocate.

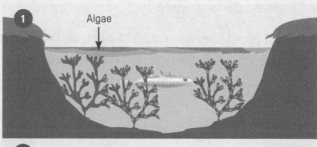

Algae

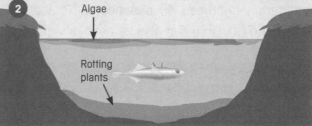

Algae

Rotting plants

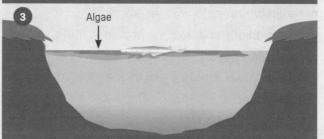

Algae

Quick Test

1. What is a fertiliser?
2. Which three elements are found in fertilisers?
3. What chemicals would you use to make ammonium sulfate in a neutralisation reaction?
4. **HT** Why do plants absorb soluble nitrate fertiliser?

Chemicals from the Sea: Sodium Chloride C2

Sodium Chloride

Sodium chloride, or table salt, is used as a food preservative and flavouring. But it is also useful as a raw material in the chemical industry. It is an important source of chlorine and sodium hydroxide.

Sodium chloride can be removed from the sea or mined from salt deposits. It is mined:
- in Cheshire as a solid (rock salt). This has led to subsidence in some parts of Cheshire
- by solution mining for the chemical industry.

Electrolysis

When concentrated sodium chloride **solution** is electrolysed, the electrodes must be made from **inert** materials as the products are very reactive. This process forms:
- sodium hydroxide in the solution.
- hydrogen at the **cathode** (negative electrode)
- chlorine at the **anode** (positive electrode). You can test for chlorine using damp litmus paper; if chlorine is present, it will bleach the litmus paper.

There are many uses for the products of **electrolysis** of sodium chloride:
- Sodium hydroxide is used to make soap.
- Hydrogen is used in the manufacture of margarine.

- Chlorine is used to sterilise water, make solvents and plastics, for example, PVC.
- Chlorine and sodium hydroxide are reacted together to make household bleach.

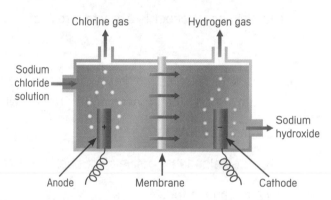

HT Electrolysis of Sodium Chloride

Brine (NaCl(aq)) contains Na^+, Cl^-, OH^- and H^+ ions. The large scale electrolysis of brine happens as part of the chloro-alkali industry. This is a global market which generates great profits.

- Hydrogen is made by **reduction** at the cathode:

$$2H^+ + 2e^- \longrightarrow H_2$$

Reduction is gain of electrons.

- Chlorine is made by **oxidation** at the anode:

$$2Cl^- - 2e^- \longrightarrow Cl_2$$

Oxidation is loss of electrons.

Sodium (Na^+) and hydroxide (OH^-) ions remain in solution. This makes the third product of sodium hydroxide.

Quick Test

1. What are the three products of the electrolysis of sodium chloride solution?
2. How is household bleach made?
3. What is a use for chlorine?

C2 Exam Practice Questions

1 The Earth is made up of three layers: the crust, the mantle and the core. What is the lithosphere? **[1]**

2 Decide whether each of the following building materials is manufactured or natural. **[3]**

Slate

Steel

Cement

Marble

Brick

3 Copper is an important metal found in ores.

a) Explain how copper is extracted from its ore. **[1]**

b) Electrical wires are made out of copper. The copper needs to be very pure. Copper is purified by electrolysis. In the process of electrolysis, what is:

i) the electrolyte?

ii) the cathode (negative electrode)?

iii) the anode (positive electrode)? **[3]**

4 Iron in the form of steel is often used to make car bodies. What two substances must be present to make iron rust? **[2]**

5 Since 2006 at least 85% of a car has to be able to be recycled.

What are the two main environmental reasons for recycling a car at the end of its useful life? **[2]**

6 Fertilisers are used by many farmers across the UK.

a) Why do farmers add fertilisers to their soil? **[1]**

b) Explain how a fertiliser works. **[2]**

7 The Haber process is used in industry to make ammonia (NH_3). The graph on the right shows how the yield changes with temperature and pressure.

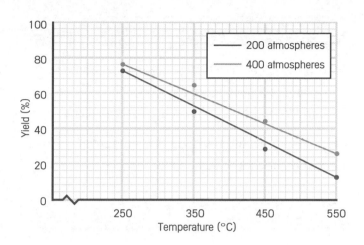

a) Name the two reactants used for making ammonia. **[1]**

b) What is the yield of ammonia when the temperature is 250°C and the pressure is 200 atmospheres? **[1]**

c) At what temperature was the yield 44% when the pressure was 400 atmospheres? **[1]**

d) If the temperature was 250°C, which pressure would give the greatest yield? **[1]**

e) Explain what happens to the yield as temperature increases. **[1]**

8 Ammonium nitrate is an example of a fertiliser that can be made from a neutralisation reaction.

a) What alkali would be reacted with nitric acid to make this fertiliser? **[1]**

HT

b) Write a balanced symbol equation for this reaction. **[3]**

9 The crust of the Earth is split into large pieces of rock known as tectonic plates.

There are different types of plate boundaries where the plates meet.

What happens at a subduction zone? **[3]**

C3 Rate of Reaction (1)

Collision Theory

Reactions happen at different speeds, for example:
- rusting is a slow reaction
- burning and an explosion are fast reactions.

The rate of reaction measures the amount of product made in a specific time.

The rate of reaction can be measured in:
- g/s or g/min for mass changes
- cm^3/s or cm^3/min for volume changes.

The rate of a reaction can be increased by:
- raising the temperature
- increasing the concentration (of liquids)
- increasing the pressure (of gases).

Rate of Reaction Experiments

Chemical reactions **stop** when one of the **reactants** is **used up**. The amount of **product formed** depends on the amount of reactant **used**.

Often there is not the same amount of each type of reactant. The limiting reactant is the one that is used up by the end of the reaction.

HT If there are more reactants, there are more reactant particles so more product particles can be made. The limiting reactant determines the maximum amount of product that can be made.

You can measure the rate of a reaction by monitoring the mass of a reaction mixture in a flask on a top pan balance or measuring the volume of a gas produced using a gas syringe.

Analysing the Rate of Reactions

From a graph, you can find out the following:
1. How long it takes to make the maximum amount of products by drawing a vertical line down to the x-axis (time) from the flat line. (The flat line on the graph shows that the reaction is finished.)
2. How much product was made by drawing a horizontal line from the highest point on the graph across to the y-axis.
3. Which reaction is quicker by comparing the steepness of the lines.

HT The **rate of reaction** can be calculated from the **slope** of a **graph** (gradient). From this you can **extrapolate** (calculate further data) and **interpolate** (create missing data). For example, you can:
1. Calculate the initial rate of reaction by drawing a straight line following the start of the curve. Now calculate the gradient and this is the rate of reaction.
2. Work out the amount of product formed for a time for which you don't have a reading.
3. Extend the curve to estimate its likely path.

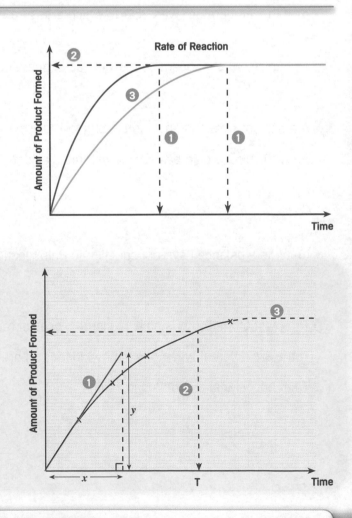

Reactant • Product

Temperature of the Reactants

Chemical reactions happen when **particles collide** with **enough energy**. The **more collisions** there are between particles, the **faster** the **reaction**.

In a reaction at **low temperature**, the particles move slowly. This means that the particles collide less often, and at lower energy, so fewer collisions will be successful. The rate of reaction will be slow.

In a reaction at **high temperature**, the particles move fast. This means that particles collide more often, and at higher energy. The rate of reaction will be fast.

Low Temperature

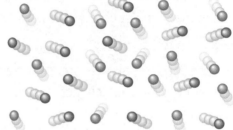

High Temperature

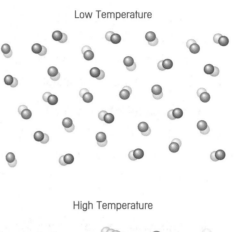

HT Increasing Temperature

Increasing temperature causes an increase in the **kinetic energy** of the particles, i.e. they move a lot faster.

The faster the particles move, the greater the chance of them colliding, so the number of collisions per second increases:

- More frequent collisions between particles lead to a faster reaction.

When the particles collide at an increased temperature they have more energy. High energy collisions increase the chance of it being a successful collision:

- More energetic collisions lead to more successful collisions.

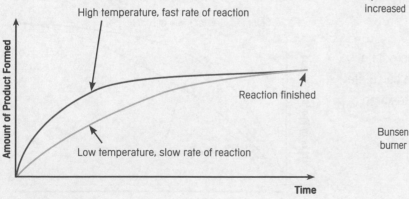

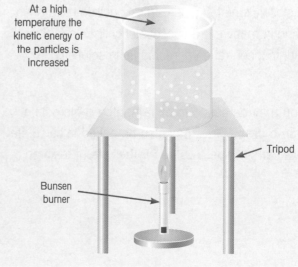

At a high temperature the kinetic energy of the particles is increased

Tripod

Bunsen burner

C3 Rate of Reaction (2)

Concentration of the Reactants

In a reaction where one or both reactants are in **low concentrations**, the particles will be spread out. The particles will collide with each other less often, so there will be fewer successful collisions.

If there is a **high concentration** of one or both reactants, the particles will be crowded close together. The particles will collide with each other more often, so there will be many more successful collisions.

(HT) Increasing concentration increases the number of particles in the same space, i.e. the particles are much more crowded together.

The more crowded the particles are, the greater the chance of them colliding, which increases the number of collisions per second:
* More frequent collisions lead to a faster reaction.

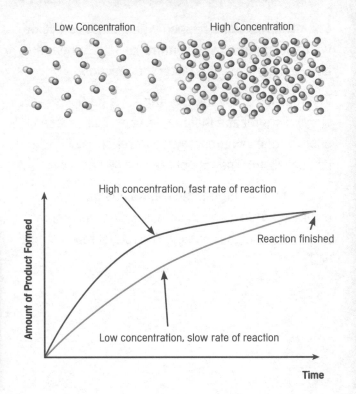

Low Concentration   High Concentration

High concentration, fast rate of reaction

Reaction finished

Amount of Product Formed

Low concentration, slow rate of reaction

Time

Pressure of a Gas

When a gas is under a **low pressure**, the particles are spread out. The particles will collide with each other less often so there will be fewer successful collisions. (This is like low concentration of reactants in solution.)

When the **pressure is high**, the particles are crowded more closely together. The particles collide more often, resulting in many more successful collisions. (This is like high concentration of reactants in solution.)

(HT) You may be asked to extrapolate or interpolate data on a graph. You may also be asked to calculate the gradient to give a value for the rate of reaction.

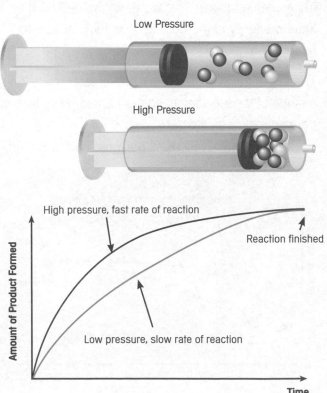

Low Pressure

High Pressure

High pressure, fast rate of reaction

Reaction finished

Amount of Product Formed

Low pressure, slow rate of reaction

Time

Concentration • Pressure

Surface Area of Solid Reactants

The **larger** the **surface area** of a reactant, the **faster** the reaction. **Powdered solids** have a larger surface area compared to their volume than **lumps of solid**. This means there are **more particles** on the **surface** for the other reactants to collide with.

The greater the number of particles exposed, the greater the chance of them colliding, which **increases** the rate of the **reaction**. So, powders can have very fast reactions, much faster than a lump of the same reactant.

HT A greater proportion of particles exposed in a powdered solid means the particles have a **greater chance** of **colliding**, which means the frequency of collisions increases. (There are more collisions each second.)

An explosion is a very fast reaction where huge volumes of gas are made.

Workers in factories that handle powders, such as flour, custard powder or sulfur, have to be very careful because the **dust** of these materials can **mix** with **air** and could cause an **explosion** if there is a **spark**.

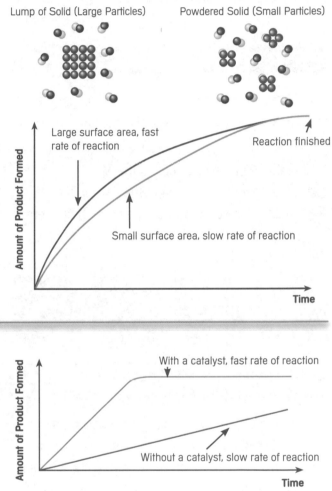

Lump of Solid (Large Particles)   Powdered Solid (Small Particles)

Large surface area, fast rate of reaction

Reaction finished

Amount of Product Formed

Small surface area, slow rate of reaction

Time

Using a Catalyst

A **catalyst** is a substance that **changes** the rate of a chemical **reaction** without being used up or changed at the end of the reaction. Catalysts are often used to speed up the rate of reaction.

Catalysts are very useful materials, as only a **small amount** of catalyst is needed to speed up the reaction of large amounts of reactant.

You can see how a catalyst affects the rate of reaction by looking at a graph. This graph shows two reactions that eventually produce the same amount of product. One reaction takes place much faster than the other because a catalyst is used.

With a catalyst, fast rate of reaction

Amount of Product Formed

Without a catalyst, slow rate of reaction

Time

HT You may be asked to extrapolate or interpolate data on a graph. You may also be asked to calculate the gradient to give a value for the rate of reaction.

Quick Test

1. What is rate of reaction?
2. Give an example of units which could be used to measure the rate of reaction for mass change reactions.
3. How can rate of reaction be increased?
4. Explain how temperature can increase the rate of reaction.

C3 Reacting Masses

Relative Atomic Mass, A_r

Every element has its own **relative atomic mass**, A_r.

Each element in the periodic table has two numbers. The **larger** of the two numbers is the A_r, for example the A_r of carbon is 12.

Relative atomic mass

12
6 **C**
carbon

Relative Formula Mass, M_r

The **relative formula mass**, M_r, of a compound is the sum of the relative atomic masses of all the atoms present in the formula.

Follow these steps to calculate the M_r:

1. Write down the formula of the compound.
2. Multiply the number of atoms of each element in the formula by its A_r.
3. Add them all up.

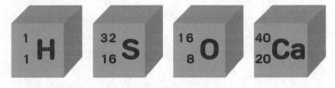

Example 1
Calculate the relative formula mass of H_2SO_4.

Write the symbols | Multiply the number of atoms in the formula by the A_r

H	2 × 1	= 2
S	1 × 32	= 32
O	4 × 16	= 64
	M_r =	**98**

Add them all up

Example 2
Calculate the relative formula mass of $Ca(OH)_2$

Ca	1 × 40	= 40
O	2 × 16	= 32
H	2 × 1	= 2
	M_r =	**74**

Calculating Reactants and Products

To calculate how much substance a reaction will produce (**product**), or the amount of starting materials (**reactant**) you need, you must remember the following:

- The total mass of the reactants always equals the total mass of the products as no atoms are created or destroyed in a chemical reaction.
- The more reactants you start with, the greater the amount of product formed.

Substances react in simple **ratios**. You can use the ratio to calculate how much of each reactant is needed to produce a certain amount of product.

The mass of the product is directly proportional to the mass of the reactant.

Sometimes you know the starting mass of only one reactant and the final mass of the product.

You can use this method to work out the missing mass.

Example 1
4.8g of magnesium reacts with oxygen to make 8.0g of magnesium oxide. What mass of oxygen will be used?

magnesium	**+**	oxygen	→	magnesium oxide
4.8g	**+**	oxygen	→	8.0g

Mass of oxygen = 8.0 − 4.8 = 3.2g

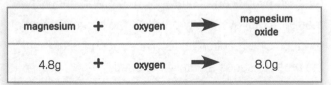

HT
$$2Mg + O_2 → 2MgO$$
2 × 24 = 48g   2 × 16 = 32g   2 × (24 + 16) = 80g

If 48g of magnesium needs 32g of oxygen but we have only 4.8g then we need 3.2g of oxygen.

Calculating Reactants and Products (Cont.)

Example 2

Calcium carbonate decomposes to make calcium oxide and carbon dioxide in the mass ratio 100 : 56 : 44. The ratio is made of the M_r for each reactant or product.

Calcium carbonate **→** Calcium oxide + Carbon dioxide
100g **→** 56g + 44g

a) Calculate how much of the reactant you would need to make 56kg of calcium oxide.

> 1kg = 1000g, so multiply all the quantities by 1000 to get the amounts in kg.

Calcium carbonate **→** Calcium oxide + Carbon dioxide
100kg **→** **56kg** + 44kg

b) Calculate how much calcium carbonate you would need to make 4g of carbon dioxide.

> $4g = \dfrac{44g}{11}$ so divide all the quantities by 11

Calcium carbonate **→** Calcium oxide + Carbon dioxide

$\dfrac{100g}{11} = \textbf{9.09g}$ **→** $\dfrac{56g}{11} = 5.1g$ + $\dfrac{44g}{11} = 4g$

HT Calculating Masses

The **total mass** of **reactants equals** the **total mass** of **products** because **no atoms** are **gained** or **lost** in a chemical reaction. There is exactly the same number of atoms; they are just rearranged into different substances.

To work out how much of a substance is used up or produced use this method:

1. Write down the balanced symbol equation for the reaction.
2. Work out the M_r of each substance.
3. Check that the total mass of reactants equals the total mass of the products.
4. Ignore the substances not mentioned in the question, and create a ratio of mass of reactant to mass of product for the substances that are mentioned.

5. Use the ratio to calculate how much of the named substance can be produced or is needed for the reaction.

Example

1. HNO_3 + NH_3 **→** NH_4NO_3
2. $1 + 14 + (3 \times 16)$ + $14 + (3 \times 1)$ **→** $14 + (1 \times 4) + 14 + (3 \times 16)$
3. 63 + 17 **→** 80
4. 63 : 17 : 80

5. If 63kg of nitric acid and 17kg of ammonia produces 80kg of ammonium nitrate then 1kg of nitric acid and 0.3kg of ammonia would produce:

$\dfrac{80}{63}$ = **1.3kg of ammonium nitrate**

Quick Test

1. What is relative atomic mass (A_r)?
2. What is relative formula mass (M_r)?

C3 Percentage Yield and Atom Economy

Percentage Yield

Percentage yield is a way of comparing the amount of product made (**actual yield**) to the amount of product expected to be made (**predicted yield**).

You can calculate percentage yield by using this formula:

$$\text{Percentage yield} = \frac{\text{Actual yield}}{\text{Predicted yield}} \times 100$$

- A 100% yield means that no product has been lost (actual yield is the same as predicted yield).
- A 0% yield means that no product has been made (actual yield is zero).

Example

A reaction was expected to produce a mass of 10g. But, the actual mass produced was 8g. Calculate the percentage yield.

$$\text{Percentage yield} = \frac{\text{Actual yield}}{\text{Predicted yield}} \times 100$$

$$= \frac{8g}{10g} \times 100 = \textbf{80\%}$$

There are several reasons why the percentage yield is less than the expected yield. The products could be lost in **evaporation**, **filtration** or during the **transfer of liquids**. Also not all reactants may have been used to make the products.

HT In industrial processes a high percentage yield is desired. This reduces waste and therefore cost.

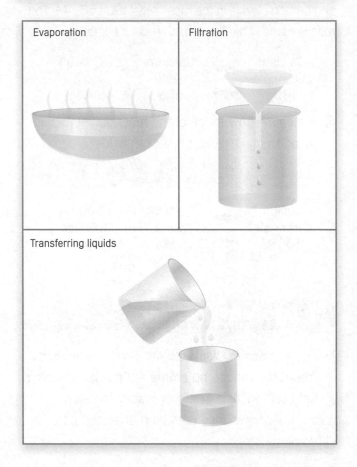

Evaporation

Filtration

Transferring liquids

Atom Economy

Atom economy is a way of measuring the number of atoms that are wasted in a chemical reaction. If all the atoms in the reactants are in the product, then the atom economy is 100%.

Example

In the hydration of ethene to make ethanol:

ethene	+	water	→	ethanol
HT C_2H_4	+	H_2O	→	C_2H_5OH

Atom economy can be calculated by using the following equation:

$$\text{Atom economy} = \frac{M_r \text{ of desired products} \times 100}{\text{Sum of } M_r \text{ of all the products}}$$

All of the atoms in the reactants are used in the product.

So, atom economy $= \dfrac{46 \times 100}{46} = 100\%$

The higher the atom economy, the 'greener' the chemical reaction.

HT In an industrial process a high atom economy is wanted as this reduces the amount of unwanted products and makes the process more sustainable.

In your exam you may be asked to look at a balanced symbol equation and say why a process has 100% atom economy or less than 100% atom economy.

Exothermic and Endothermic Reactions

Chemical reactions either **give out** or **take in** energy. So chemical reactions cause a temperature change.

Exothermic reactions release energy to the surroundings and cause a temperature rise.

Endothermic reactions absorb energy from the surroundings and cause a temperature drop.

The energy given out by exothermic chemical reactions can be used to:
* heat things
* produce electricity
* make sound
* make light.

Energy is measured in **joules (J)** or **kilojoules (kJ)**.

Temperature is measured in **degrees Celsius (°C)**.

Comparing Fuels

A **calorimeter** can be used to compare the amounts of heat energy released by the combustion of different fuels. This is a **calorimetry** experiment.

If you burn the same mass of each fuel, the fuel that produces the largest temperature rise releases the most energy.

The formula used to work out the change in temperature (°C) is:

$$\text{Temperature change} = \text{Final temperature of water (°C)} - \text{Start temperature of water (°C)}$$

To make meaningful comparisons (i.e. to do a **fair test**) you need to:
* use the same mass (or volume) of water
* use the same calorimeter
* have the burner and calorimeter the same distance apart
* burn the same mass of fuel.

To make your calorimetry reliable you should repeat your experiment and take an average (mean) of the temperature rise. To help you compare between fuels, you may wish to calculate the energy released per gram of fuel. To do this you would need to measure the mass of the spirit burner before and after the experiment.

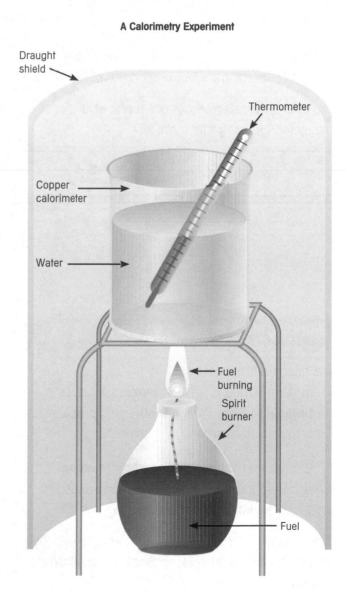

A Calorimetry Experiment

Draught shield

Thermometer

Copper calorimeter

Water

Fuel burning

Spirit burner

Fuel

C3 Energy

Calculating Energy Changes

To compare fuels, you need to work out the amount of energy transferred by the fuel to the water in the **calorimetry** experiment. The amount of energy transferred can be calculated by using the following formula:

Energy transferred (J)	=	Mass of water heated (g)	×	Specific heat capacity (J/g/°C)	×	Change in temperature (°C)
q	=	M	×	C	×	ΔT

N.B. Specific heat capacity is a constant that is specific to a particular material. Water has a value of 4.2J/g/°C.

HT To compare fuels you need to work out the amount of energy transferred per gram of fuel burned.

The energy transferred per gram of fuel is calculated by using this formula:

$$\text{Energy per gram} = \frac{\text{Energy supplied (J)}}{\text{Mass of fuel burned (g)}}$$

Making and Breaking Bonds

In a chemical reaction:
- **making** bonds is an **exothermic** process
- **breaking** bonds is an **endothermic** process.

HT Chemical reactions that need more energy to break bonds than released when new bonds are made are **endothermic reactions**.

Chemical reactions that release more energy when making bonds than breaking them are **exothermic reactions**.

Example

The results from a calorimeter experiment with hexane are as follows:
- Mass of hexane burned = 0.26g
- Rise in temperature of water = 12°C
- Mass of water in calorimeter = 200g

Calculate the energy transferred.

Energy transferred (J)	=	Mass of water heated (g)	×	Specific heat capacity (J/g/°C)	×	Change in temperature (°C)

Energy transferred = 200g × 4.2J/g/°C × 12°C

= **10 080 joules**

Example

Calculate the amount of energy per gram of hexane fuel.

$$\text{Energy per gram} = \frac{\text{Energy supplied (J)}}{\text{Mass of fuel burned (g)}}$$

$$\text{Energy per gram} = \frac{10\,080J}{0.26g}$$

= **38 769J/g**

N.B. The actual value for hexane is 48 407J/g. Our result is lower because some energy is lost to the surroundings and some is transferred to the calorimeter.

Quick Test

1. When 12g of carbon is completely burned in oxygen, 10g of carbon dioxide is made. Calculate the percentage yield.
2. What is the formula for atom economy?
3. In the blue Bunsen flame, methane burns with lots of oxygen to make carbon dioxide and water only.
   a) Write a word equation for this reaction.
   b) Is this reaction exothermic or endothermic?

Batch and Continuous Processes

In a **batch process**, reactants are put into a reactor and the product is removed at the end of the reaction. Medicines and pharmaceutical drugs can be made in this way. Batch processes:
- make a product on demand and on a small scale
- can be used to make a variety of products
- are labour intensive – the reactor needs to be filled, emptied and cleaned.

In a **continuous process,** reactants are continually fed into the reactor as the products are removed. The production of ammonia in the Haber process and sulfuric acid in the Contact process are made in this way. Continuous processes:
- make a product on a large scale
- are dedicated to making just one product
- operate all the time
- can run automatically.

Making and Developing Medicines

The materials used to make a medicine can be extracted from plant materials by:
- **crushing** – using a pestle and mortar
- **boiling and dissolving** – using a suitable solvent
- **chromatography** – using chromatography to separate a concentrated solution.

Developing a new pharmaceutical drug is very expensive and takes a long time. The drug must also be approved for use, and satisfy all the legal requirements set out by the government.

The costs include:
- **materials** needed – they could be synthetic or natural
- **research** development and **testing** which can take many years
- **labour** – highly qualified staff are needed and often lots of staff are needed. It can't be automated (carried out by machines) as only small quantities are made
- **energy**
- **marketing**.

It is important that pharmaceutical drugs are as pure as possible. **Thin layer chromatography** (TLC) can be used to test the purity of a chemical. Testing the melting and boiling point can help to identify the compound and infer its purity.

HT Research and development can take a few years, but it can take even longer to carry out **safety tests**, including testing on human volunteers. There are very strict legal rules which a new medicine must satisfy before it can be put on the market.

A pharmaceutical company may invest hundreds of millions of pounds to develop one drug. But they have a limited time to recoup their investment before they lose exclusive rights to make the medicine. If the number of people using the medicine is small, then the cost of buying the drug would be very high.

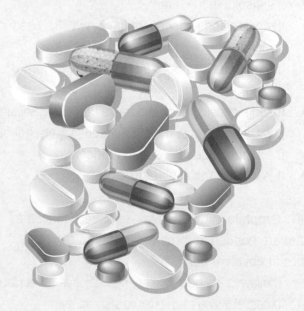

C3 Allotropes of Carbon and Nanochemistry

Carbon

There are three forms of carbon you need to know:
- **diamond**
- **graphite**
- **buckminster fullerene** (buckyballs).

These are all **allotropes** of carbon. Allotropes are different forms of the same element with atoms arranged in different structures.

All of these substances are made only of carbon atoms and have different structures.

Diamond

Diamond has a rigid structure:
- It's insoluble in water and doesn't conduct electricity.
- It's used in jewellery because it's colourless, clear (transparent) and lustrous (shiny).
- It can be used in cutting tools because it's very hard and has a very high melting point.
- It is a giant molecular structure.

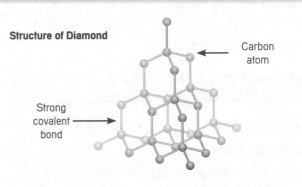

Structure of Diamond

Carbon atom

Strong covalent bond

HT **Diamond** is made of carbon atoms bonded to four other carbon atoms by strong **covalent bonds**.
- It doesn't have any free electrons so it doesn't conduct electricity.

- It's hard and has a high melting point because of the large number of covalent bonds These bonds need a lot of energy to break.

Graphite

Graphite has a layered structure:
- It's insoluble in water.
- It's black and slippery, which is why it's used in pencil leads.
- It's lustrous and opaque (light can't travel through it).
- It conducts electricity and has a very high melting point, so is used to make electrodes for electrolysis.
- It's slippery, so it's used in lubricants.
- It is a giant molecular structure.

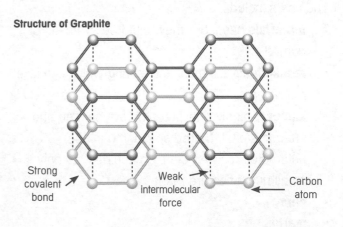

Structure of Graphite

Strong covalent bond

Weak intermolecular force

Carbon atom

HT **Graphite** is made of layers of carbon atoms that are bonded to three other carbon atoms by strong **covalent bonds**.
- The layers are held together by weak intermolecular forces, allowing each layer to slide easily.

- It conducts electricity because it has **free** (delocalised) **electrons**.
- It has a high melting point because it has many strong covalent bonds to break. These bonds need a lot of energy to break.

Allotropes of Carbon and Nanochemistry C3

Buckminster Fullerene

A **Buckminster Fullerene** molecule (C_{60}) – known as a 'buckyball' – is made of 60 carbon atoms arranged in a football-like sphere. It is a black solid.

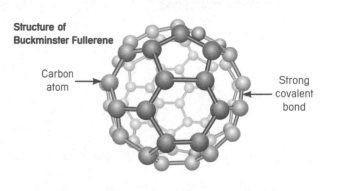

Structure of Buckminster Fullerene

Carbon atom

Strong covalent bond

Nanochemistry and Nanotubes

Chemistry deals with materials on a **large scale**, but **nanochemistry** deals with materials on an **atomic scale** (i.e. individual atoms).

Chemists discovered that **nanotubes** could be made by joining fullerenes together. Nanotubes conduct electricity and are very strong. They are used to:

- reinforce graphite tennis racquets because of their strength
- make connectors and semiconductors in circuits because of their electrical properties
- develop more efficient industrial catalysts.

Fullerenes and nanotubes can be used to 'cage' other molecules because their shape allows them to trap other substances, for example:

- **drugs**, e.g. a major new HIV treatment uses buckyballs to deliver a material which disrupts the working of the HIV virus

HT • **catalysts** – by attaching catalyst material to a nanotube, a massive surface area can be achieved, making the catalyst very efficient.

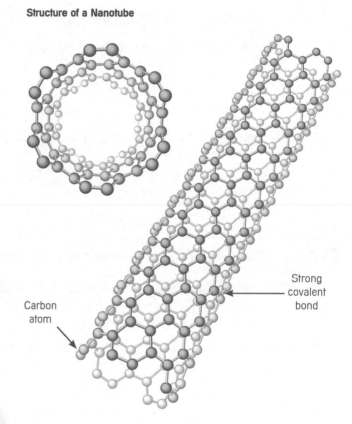

Structure of a Nanotube

Carbon atom

Strong covalent bond

Quick Test

1. Give an example of a chemical that is made by batch production.
2. What are the four stages for extracting chemicals from plants?
3. Name three allotropes of carbon.
4. What is the formula of buckminster fullerene?

C3 Exam Practice Questions

1 a) Using the periodic table to help you, calculate the relative formula mass of each of the following compounds.

    i) Na_2CO_3 **[2]**

    ii) $(NH_4)_2SO_4$ **[3]**

b) A neutralisation reaction was expected to produce a salt with a mass of 15g. The actual mass was 9g.

Calculate the percentage yield.   **[2]**

2 a) Pharmaceutical drugs are made using a batch process.

Explain why the materials needed to make new medicines are expensive.   **[1]**

b) Describe the advantages of using a batch process to make a chemical.   **[3]**

3 Collision theory can be used to explain how changing conditions affects the rate of reaction.

    a) What must happen to particles in order for a reaction to take place?   **[2]**

    b) What happens to particles when you increase the concentration?   **[2]**

    c) How does increasing the temperature increase the rate of reaction?   **[2]**

4 When powdered calcium carbonate is put into an acid it reacts quicker than if a lump of calcium carbonate was used.

Explain why the powdered solid reacts faster than a lump of the same substance.   **[3]**

5 When magnesium metal is put into a beaker of hydrochloric acid a chemical reaction happens. Chantal completed an experiment to measure the gas made in the reaction. She plotted her results on the graph to the right.

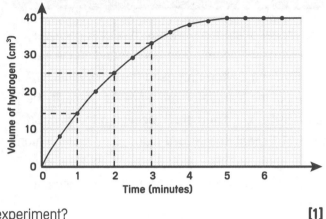

The Reaction between Magnesium and Dilute Hydrochloric Acid

a) What was the independent variable in Chantal's experiment? **[1]**

_____

b) What was the maximum volume of gas made in this experiment? **[1]**

_____

c) When did the reaction stop? **[1]**

_____

6 Rate of reaction can be measured by monitoring the amount of product made over time. This data can then be put on a graph.

a) Write down two factors that affect the rate of reaction. **[2]**

_____

_____

HT

b) How can you use a graph to calculate the rate of reaction? **[1]**

_____

7 Iron combusts in oxygen to make iron oxide (Fe_2O_3).

a) Write a balanced symbol equation for this reaction. **[3]**

_____

b) Calculate the relative formula mass (M_r) for iron oxide. (A_r of Fe = 56, A_r of O = 16) **[2]**

_____

c) Mass is conserved in this chemical reaction. Use relative formula mass and atomic masses to show this. **[4]**

_____

_____

C4 Atomic Structure

History of the Atom

The model of atomic structure has changed over time as new evidence has been found. In the early 1800s **John Dalton** proposed the theory that all atoms of the same element were the same. In the late 1890s **J. J. Thomson** discovered the electron. In 1911 **Ernest Rutherford** discovered that the atom had a dense centre called the nucleus. Then in 1913 Niels Bohr predicted that electrons occupy orbitals.

It is the work of all these scientists that have led to the current theory of atomic structure.

(HT) Some unexpected results from scientists like Geiger and Marsden have led to the model of the atom being modified in order to explain them.

Structure of an Atom

An **atom** has a central **nucleus** surrounded by shells of negatively charged **electrons**.

The nucleus is made up of **protons** and **neutrons**. The nucleus is positively charged but the atom has no overall charge.

(HT) An atom has no overall charge because it has the **same number** of (positive) **protons** and (negative) **electrons**. So, the charges **cancel** each other out.

Atoms have a radius of about 1^{-10}m and a mass of about 10^{-23}g.

Atomic Particle	Relative Charge	Relative Mass
Proton	+1	1
Neutron	0	1
Electron	−1	0.0005 (zero)

Structure of a Fluorine Atom

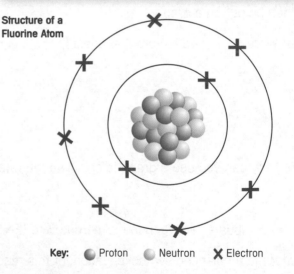

Key: ● Proton ● Neutron ✗ Electron

Elements and Compounds

An **element** is made of one type of atom. Elements can't be chemically broken down. There are just over 100 naturally occurring elements.

A **compound** is a substance made of two or more elements that are **chemically combined**. You can identify the elements in a compound from its formula, using the periodic table.

For example:
- the **compound** sodium chloride (NaCl) contains the **elements** sodium (Na) and chlorine (Cl)
- the **compound** potassium nitrate (KNO_3) contains the **elements** potassium (K), nitrogen (N) and oxygen (O).

Key Words   Atom • Nucleus • Electron • Proton • Neutron • Element • Compound

Mass Number and Atomic Number

The **mass number** is the total number of **protons** and **neutrons** in an atom.

The **atomic number** (proton number) is the number of **protons** in an atom.

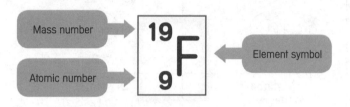

The elements in the periodic table are arranged in **increasing atomic number**.

You can use the periodic table to find:
* an element from its atomic number
* the atomic number of an element.

The group number is the same as the number of electrons in the outer shell of an element's atom. The period number is the same as the number of occupied shells (i.e. shells that contain electrons) that an element's atom has.

You can work out the number of protons, electrons and neutrons in an atom or ion if you know its atomic number, mass number and charge.

Element Symbol	Protons	Electrons	Neutrons
Hydrogen atom $_{1}^{1}H$	1	1	0 $1 - 1 = 0$
Helium atom $_{2}^{4}He$	2	2	2 $4 - 2 = 2$
Sodium atom $_{11}^{23}Na$	11	11	12 $23 - 11 = 12$
(HT) Oxygen ion $_{8}^{16}O^{2-}$	8	$8 + 2 = 10$	8 $16 - 8 = 8$

Isotopes

Isotopes are **atoms** of the **same element** that have the **same atomic number** but a **different mass number**.

For example, chlorine has two isotopes:

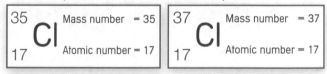

(HT) You can identify isotopes from data about the number of electrons, protons and neutrons in particles. Each isotope has the **same number of** **protons** and **electrons**, but a **different number** of **neutrons**. For example, carbon has three main isotopes, as listed in this table:

Isotope	Symbol	Mass Number	Atomic Number	Protons	Neutrons	Electrons
Carbon-12	$_{6}^{12}C$	12	6	6	6	6
Carbon-13	$_{6}^{13}C$	13	6	6	7	6
Carbon-14	$_{6}^{14}C$	14	6	6	8	6

C4 Atomic Structure

Electron Configuration

Electron configuration tells you how the electrons are arranged around the nucleus in **shells** (**energy levels**):

The first shell can hold a maximum of two electrons, and the second and third hold a maximum of eight electrons.

Hydrogen, H Atomic No. = 1	Helium, He Atomic No. = 2	Lithium, Li Atomic No. = 3	Beryllium, Be Atomic No. = 4
1	2	2.1	2.2
Boron, B Atomic No. = 5	**Carbon, C** Atomic No. = 6	**Nitrogen, N** Atomic No. = 7	**Oxygen, O** Atomic No. = 8
2.3	2.4	2.5	2.6
Fluorine, F Atomic No. = 9	**Neon, Ne** Atomic No. = 10	**Sodium, Na** Atomic No. = 11	**Magnesium, Mg** Atomic No. = 12
2.7	2.8	2.8.1	2.8.2
Aluminium, Al Atomic No. = 13	**Silicon, Si** Atomic No. = 14	**Phosphorus, P** Atomic No. = 15	**Sulfur, S** Atomic No. = 16
2.8.3	2.8.4	2.8.5	2.8.6
Chlorine, Cl Atomic No. = 17	**Argon, Ar** Atomic No. = 18	**Potassium, K** Atomic No. = 19	**Calcium, Ca** Atomic No. = 20
2.8.7	2.8.8	2.8.8.1	2.8.8.2

Quick Test

1. What particles are found in the nucleus of an atom?
2. What is an element?
3. a) What is the symbol for fluorine?
   b) What is the mass number of fluorine?
   c) What is the atomic number of fluorine?
   d) How many neutrons does an atom of fluorine have?

Ions

An **ion** is a **charged atom** or group of atoms, e.g. Na^+, Cl^-, NH_4^+, SO_4^{2-}.

A **positive ion** is made when an atom, or group of atoms, **loses** one or more **electrons**. For example, losing two electrons makes a 2^+ ion, e.g. Mg^{2+}.

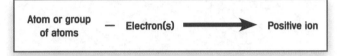

A **negative ion** is made when an atom, or group of atoms, **gains** one or more electrons. For example, gaining two electrons makes a 2^- ion, e.g. O^{2-}.

| Atom or group of atoms | + | Electron(s) | ⟶ | Negative ion |

Ionic Bonding

In **ionic bonding**:

- the metal atom loses all outer-shell **electrons** to become a **positive ion**
- the non-metal atom **gains electrons** to fill its outer shell and become a **negative ion**
- the positive and negative ions are attracted to each other. This attraction is an **ionic bond**.

Two ionically bonded compounds are **sodium chloride** and **magnesium oxide**. They have high melting points. They don't conduct electricity when solid.

But both of these compounds dissolve in water and **can conduct electricity** when **in solution**. They can also **conduct electricity** when **molten**.

Structure and Physical Properties

Sodium chloride (NaCl) and magnesium oxide (MgO) form **giant ionic lattices** in which positive and negative ions are **strongly attracted** to each other.

HT This means that they:

- have high melting points as there is a strong attraction between oppositely charged ions
- can conduct electricity when molten or in solution because the charged ions are free to move about
- don't conduct electricity when solid, because the ions are held in place and can't move.

But magnesium oxide has a higher melting point than sodium chloride as the ionic bonds are stronger and need more energy to be broken.

In your exam you may be asked to predict the properties of other giant ionic structures. Use your knowledge about magnesium oxide and sodium chloride to help you.

Sodium Chloride

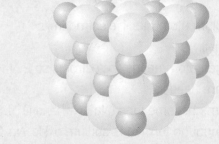

Na^+ ion, i.e. a sodium atom that has lost 1 electron
Cl^- ion, i.e. a chlorine atom that has gained 1 electron

C4 Ionic Bonding

The Ionic Bond

When a metal and a non-metal combine, electrons are transferred from one **atom** to the other forming **ions**. Each ion will have a complete outer shell (**a stable octet**).

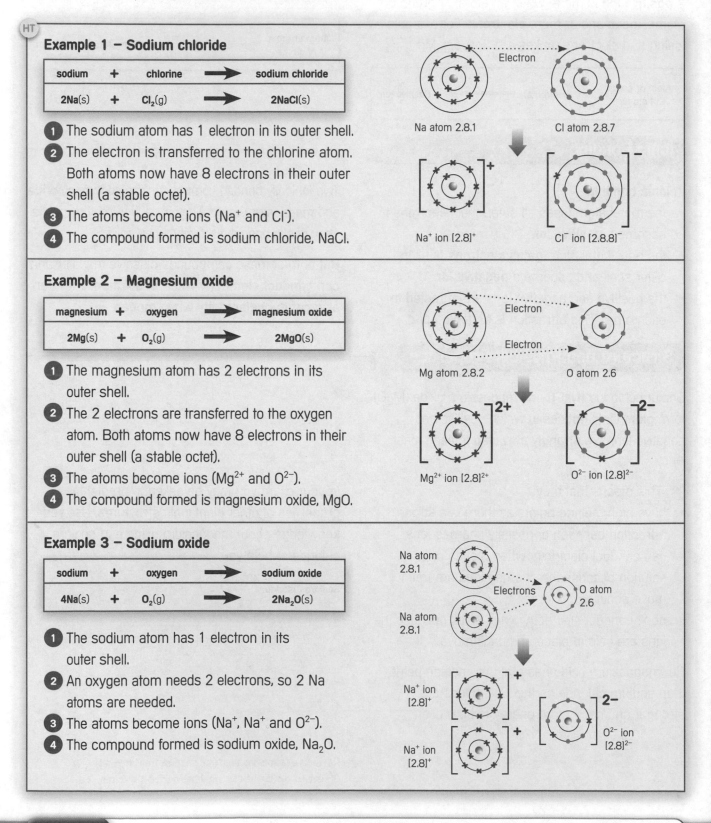

HT

Example 1 – Sodium chloride

sodium	+	chlorine	⟶	sodium chloride
2Na(s)	+	**Cl₂**(g)	⟶	**2NaCl**(s)

1. The sodium atom has 1 electron in its outer shell.
2. The electron is transferred to the chlorine atom. Both atoms now have 8 electrons in their outer shell (a stable octet).
3. The atoms become ions (Na^+ and Cl^-).
4. The compound formed is sodium chloride, NaCl.

Na atom 2.8.1   Cl atom 2.8.7
Na^+ ion $[2.8]^+$ Cl^- ion $[2.8.8]^-$

Example 2 – Magnesium oxide

magnesium	+	oxygen	⟶	magnesium oxide
2Mg(s)	+	**O₂**(g)	⟶	**2MgO**(s)

1. The magnesium atom has 2 electrons in its outer shell.
2. The 2 electrons are transferred to the oxygen atom. Both atoms now have 8 electrons in their outer shell (a stable octet).
3. The atoms become ions (Mg^{2+} and O^{2-}).
4. The compound formed is magnesium oxide, MgO.

Mg atom 2.8.2   O atom 2.6
Mg^{2+} ion $[2.8]^{2+}$ O^{2-} ion $[2.8]^{2-}$

Example 3 – Sodium oxide

sodium	+	oxygen	⟶	sodium oxide
4Na(s)	+	**O₂**(g)	⟶	**2Na₂O**(s)

1. The sodium atom has 1 electron in its outer shell.
2. An oxygen atom needs 2 electrons, so 2 Na atoms are needed.
3. The atoms become ions (Na^+, Na^+ and O^{2-}).
4. The compound formed is sodium oxide, Na₂O.

Na atom 2.8.1   Na atom 2.8.1   O atom 2.6
Na^+ ion $[2.8]^+$ Na^+ ion $[2.8]^+$ O^{2-} ion $[2.8]^{2-}$

HT The Ionic Bond (Cont.)

Example 4 – Magnesium chloride

magnesium	+	chlorine	$\longrightarrow$	magnesium chloride
$Mg(s)$	+	$Cl_2(g)$	$\longrightarrow$	$MgCl_2(g)$

1 The magnesium atom has 2 electrons in its outer shell.

2 A chlorine atom only needs 1 electron, so 2 Cl atoms are needed.

3 The atoms become ions (Mg^{2+}, Cl^- and Cl^-).

4 The compound formed is magnesium chloride, $MgCl_2$.

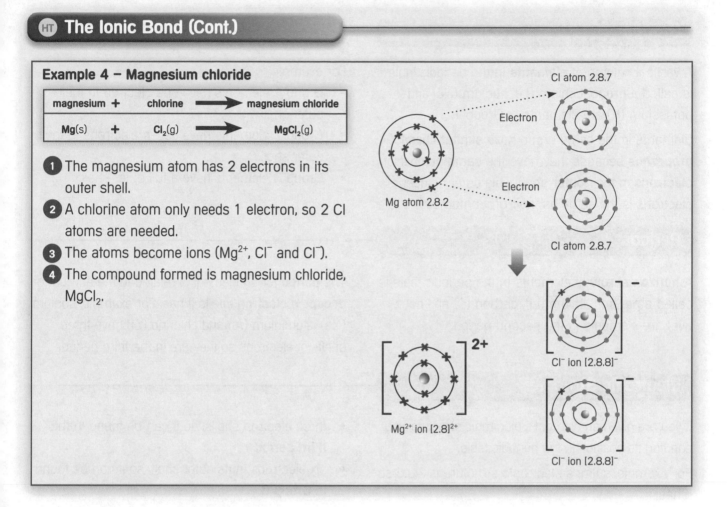

Mg atom 2.8.2

Electron

Cl atom 2.8.7

Electron

Cl atom 2.8.7

Mg^{2+} ion $[2.8]^{2+}$

Cl^- ion $[2.8.8]^-$

Cl^- ion $[2.8.8]^-$

Formulae of Ionic Compounds

Ions with different charges combine to form **ionic compounds** which have **equal amounts** of **positive** and **negative** charge.

Positive Ions							
1+ e.g. K^+, Na^+		**2+** e.g. Mg^{2+}, Cu^{2+}		**3+** e.g. Al^{3+}, Fe^{3+}			**Negative Ions**
KCl 1+ 1−	NaOH 1+ 1−	$MgCl_2$ 2+ 2 × 1− = 2−	$Cu(OH)_2$ 2+ 2 × 1− = 2−	$AlCl_3$ 3+ 3 × 1− = 3−	$Fe(OH)_3$ 3+ 3 × 1− = 3−	**1−** e.g. Cl^-, OH^-	
K_2SO_4 2 × 1+ 2− = 2+	Na_2O 2 × 1+ 2− = 2+	$MgSO_4$ 2+ 2−	CuO 2+ 2−	$Al_2(SO_4)_3$ 2 × 3+ 3 × 2− = 6+ = 6−	Fe_2O_3 2 × 3+ 3 × 2− = 6+ = 6−	**2−** e.g. SO_4^{2-}, O^{2-}	

C4 The Periodic Table and Covalent Bonding

Groups

A vertical column of **elements** in the periodic table is called a **group**. Lithium (Li), sodium (Na) and potassium (K) are elements in Group 1.

Elements in the same group have **similar chemical properties** because they have the **same number** of **electrons** in their **outer shell**. This outer number of electrons is the same as their group number.

For example:
- Group 1 elements have one electron in their outer shell.
- Group 7 elements have seven electrons in their outer shell.
- Group 0 elements have a full outer shell.

Periods

A **horizontal row** of elements in the periodic table is called a **period**. Lithium (Li), carbon (C) and neon (Ne) are elements in the second period.

The **period** for an element is related to the number of **occupied electron shells** it has. For example, sodium (Na), aluminium (Al) and chlorine (Cl) have three shells of electrons so they are in the third period.

Electronic Structure

If you are given an element's electronic structure, you can find its position in the periodic table.

For example, sulfur's electronic structure is 2.8.6 so it has:

- three electron shells, so it can be found in the **third period**
- six electrons in its outer shell, so it can be found in **Group 6**.

Bonding

There are three types of bonding:
- **Ionic bonding** between metals and non-metals.

- **Covalent bonding** between non-metals.
- Metallic bonding for metals only.

Covalent Bonding

Covalent bonding is when **non-metals** combine by **sharing pairs** of **electrons**. Water and carbon dioxide are both covalently bonded molecules. Water (H_2O) contains hydrogen and oxygen atoms. It:
- is a liquid at room temperature and has a low melting point
- doesn't conduct electricity.

One molecule of water is made up of one atom of oxygen and two atoms of hydrogen.

Carbon dioxide (CO_2) contains carbon and oxygen atoms. Carbon dioxide:

- is a gas at room temperature and has a low melting point
- doesn't conduct electricity.

One molecule of carbon dioxide is made up of **one atom of carbon** and **two atoms of oxygen**.

Simple covalently bonded molecules, e.g. water and carbon dioxide have **weak intermolecular** forces of attraction between molecules.

HT Simple covalently bonded molecules have low melting points. They don't conduct electricity because there aren't any free electrons.

Group • Period • Covalent bonding

The Periodic Table and Covalent Bonding C4

HT Representing Molecules

You should be familiar with how simple covalently bonded molecules are formed.

Hydrogen (H₂) – the two hydrogen atoms share a pair of electrons.

Chlorine (Cl₂) – the two chlorine atoms share a pair of electrons.

Methane (CH₄) – the carbon atom shares a pair of electrons with each hydrogen atom.

Carbon dioxide (CO₂) – the outer shells of the carbon and oxygen atoms overlap. The carbon atom shares two pairs of electrons with each oxygen atom to form a double covalently bonded molecule.

Water (H₂O) – the outer shells of the hydrogen and oxygen atoms overlap. The oxygen atom shares a pair of electrons with each hydrogen atom to form a water molecule.

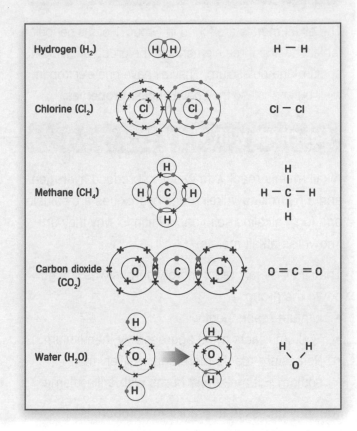

Development of the Periodic Table

Many people have researched the properties of elements and tried to classify them.

Dobereiner was the first to suggest a **Law of Triads**, where he grouped the elements into sets of three with similar properties. The middle element would have the average mass of the other two elements. But not all the elements were known and the pattern did not work for every known element.

John Newlands was the first scientist to make a table of elements, which he called the **Law of Octaves**, where every eighth element behaved the same. But he included some compounds in his table as he believed them to be elements.

Mendeleev was the author of the modern periodic table.

HT Mendeleev left gaps in his table for the unknown elements and made predictions about their properties. His predictions were later proved correct. Also, investigations on atomic structure agreed with his ideas.

Quick Test

1. Explain how sodium atoms become sodium ions.
2. What is an ionic bond?
3. Magnesium oxide is an ionic compound.
   a) When can this compound conduct electricity?
   b) What is the formula of this compound?
4. What is a covalent bond?

C4 The Group 1 Elements

Group 1 – The Alkali Metals

The **alkali metals** are found in Group 1 of the periodic table. The first three elements in the group are lithium, sodium and potassium. They all have one electron in their outer shell so they have **similar properties**.

Alkali metals are stored **under oil** because they:
- react with air
- react vigorously with water.

Reactions with Water

Alkali metals react with water to produce **hydrogen** and a **hydroxide**. Alkali metal hydroxides are soluble and form alkaline solutions, which is why they are known as **alkali** metals.

The alkali metals react more vigorously as you go down the group:
- Lithium reacts gently.
- Sodium reacts more aggressively than lithium.
- Potassium reacts more aggressively than sodium – it melts and burns with a lilac flame.

You may be asked to predict the properties of rubidium and caesium. Use your knowledge of other alkali metals to help you.

lithium + water ➡ lithium hydroxide + hydrogen

$$2Li(s) + 2H_2O(l) \longrightarrow 2LiOH(aq) + H_2(g)$$

sodium + water ➡ sodium hydroxide + hydrogen

$$2Na(s) + 2H_2O(l) \longrightarrow 2NaOH(aq) + H_2(g)$$

potassium + water ➡ potassium hydroxide + hydrogen

$$2K(s) + 2H_2O(l) \longrightarrow 2KOH(aq) + H_2(g)$$

Flame Tests

Lithium, sodium and potassium compounds can be recognised by the colours they make in a **flame test**.

1. A piece of clean nichrome wire is dipped in water.
2. The wire is dipped in the solid compound (known as the sample). The wire is then put into a Bunsen flame.
3. Each compound will produce a different coloured flame.

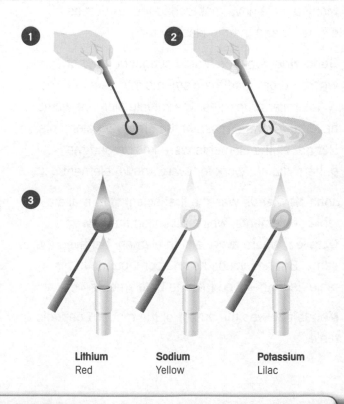

| Lithium | Sodium | Potassium |
| Red | Yellow | Lilac |

Key Words **Hydroxide**

HT Properties of the Alkali Metals

Alkali metals have similar chemical and physical properties.

Rubidium is the fourth element in Group 1. Rubidium's reaction with water is:

- very fast
- **exothermic** (gives out energy)
- violent (if it's carried out in a glass beaker, the beaker may shatter).

Density increases as you go down the group (with the exception of potassium). Caesium has the greatest density, and lowest melting and boiling points.

Element	Symbol	Melting Point (°C)	Boiling Point (°C)	Density (g/cm³)
Lithium	Li	180	1340	0.53
Sodium	Na	98	883	0.97
Potassium	K	64	760	0.86
Rubidium	Rb	39	688	1.53
Caesium	Cs	29	671	1.90

Trends in Group 1

Alkali metals have similar chemical properties because as they react, each atom **loses** one **electron** from its outer shell. So, a **positive ion** with a stable electronic structure is made.

The alkali metals become **more reactive** as you go down the group because the outer shell gets **further away** from the positive attraction of the **nucleus**. This makes it easier for an atom to lose an electron from its outer shell.

The equations for the formation of the Group 1 metal ions are:

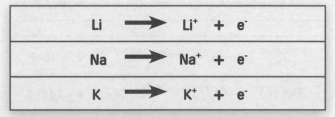

Li $\longrightarrow$ Li⁺ + e⁻

Na $\longrightarrow$ Na⁺ + e⁻

K $\longrightarrow$ K⁺ + e⁻

Oxidation involves the loss of electrons by an atom. If the ionic equation for a reaction shows that an electron has been lost, an oxidation reaction took place.

Examples of Oxidation

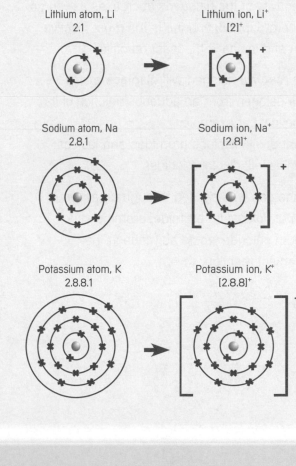

Lithium atom, Li
2.1

Lithium ion, Li⁺
[2]⁺

+ 1e⁻

Sodium atom, Na
2.8.1

Sodium ion, Na⁺
[2.8]⁺

+ 1e⁻

Potassium atom, K
2.8.8.1

Potassium ion, K⁺
[2.8.8]⁺

+ 1e⁻

Quick Test

1 Why do Group 1 metals have similar chemical properties?
2 What colour is a flame when a lithium compound is added?

C4 The Group 7 Elements

Group 7 – The Halogens

The five non-metals in Group 7 are known as the **halogens**. They all have seven electrons in their outer shell so they have similar chemical properties.

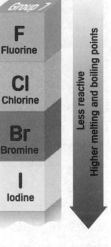

Fluorine, chlorine, bromine and iodine are halogens. At room temperature:

- chlorine is a green gas
- bromine is an orange liquid
- iodine is a grey solid.

The halogens have many uses:

- **Iodine** is used as an **antiseptic** to sterilise wounds.
- **Chlorine** is used to **sterilise water**, to make **pesticides** and to make **plastics**.

Halogens react vigorously with **alkali metals** to form metal **halides**, for example:

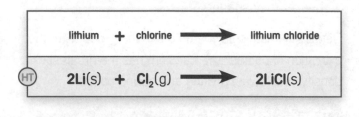

lithium	+	chlorine	⟶	lithium chloride

(HT) $2Li(s) + Cl_2(g) \longrightarrow 2LiCl(s)$

Displacement Reactions

The reactivity of the halogens decreases as you go down the group. So, fluorine is the most reactive halogen and iodine is the least reactive.

A **more reactive** halogen will **displace** a **less reactive** halogen from an aqueous solution of its metal halide. For example:

- chlorine will displace bromides and iodides
- bromine will displace iodides.

If chlorine gas was passed through an aqueous solution of potassium bromide, bromine and potassium chloride would be made in the displacement reaction.

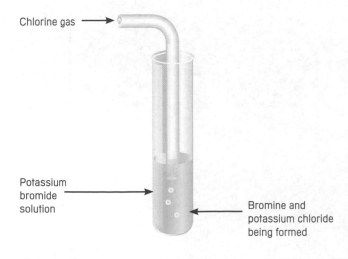

Chlorine gas

Potassium bromide solution

Bromine and potassium chloride being formed

The products of reactions between halogens and aqueous solutions of salts are as follows:

Halogen \ Halide salt	Potassium Chloride, KCl	Potassium Bromide, KBr	Potassium Iodide, KI
Chlorine, Cl_2	No reaction	Potassium chloride + bromine	Potassium chloride + iodine
Bromine, Br_2	No reaction	No reaction	Potassium bromide + iodine
Iodine, I_2	No reaction	No reaction	No reaction

potassium bromide	+	chlorine	⟶	potassium chloride	+	bromine

(HT) $2KBr(aq) + Cl_2(g) \longrightarrow 2KCl(aq) + Br_2(aq)$

potassium iodide	+	chlorine	⟶	potassium chloride	+	iodine

(HT) $2KI(aq) + Cl_2(g) \longrightarrow 2KCl(aq) + I_2(aq)$

potassium iodide	+	bromine	⟶	potassium bromide	+	iodine

(HT) $2KI(aq) + Br_2(l) \longrightarrow 2KBr(aq) + I_2(aq)$

You make be asked to suggest if a displacement reaction will happen. Remember that chlorine is more reactive than bromine, which is more reactive than iodine.

Key Words   Halogen • Halide

HT Properties of the Halogens

The physical and chemical properties of the halogens change as you go down the group. **Fluorine** is the most reactive element in the group. It will **displace** all of the other halogens from an aqueous solution of their metal halides.

Astatine is a semi-metallic, radioactive element and only very small amounts are found naturally. It's the least reactive of the halogens and, theoretically, it would be unable to displace any of the other halogens from an aqueous solution of their metal halides.

Astatine is very unstable and difficult to study. So, the information for astatine is estimated by looking at the trend in boiling point, melting point and density as you go down Group 7.

Element	Symbol	Melting Point (°C)	Boiling Point (°C)	Density (g/cm³)
Fluorine	F	−220	−188	0.0016
Chlorine	Cl	−101	−34	0.003
Bromine	Br	−7	59	3.12
Iodine	I	114	184	4.95
Astatine	At	302 (estimated)	337 (estimated)	7 (estimated)

Trends in Group 7

The halogens have similar chemical properties because, as they react, each atom **gains** one **electron** to form a **negative** ion with a stable electronic structure.

Reduction involves the gain of electrons by an atom, for example:

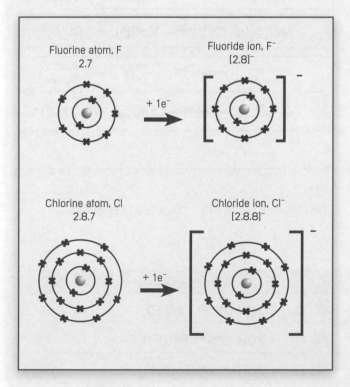

Fluorine atom, F
2.7

Fluoride ion, F⁻
[2.8]⁻

+ 1e⁻

Chlorine atom, Cl
2.8.7

Chloride ion, Cl⁻
[2.8.8]⁻

+ 1e⁻

The halogens at the top of the group are **more reactive** than those at the bottom of the group because the outer shell is **closer** to the **positive attraction** of the **nucleus**. This makes it **easier** for an atom to **gain an electron**.

Equations for the formation of the halide **ions** from halogen molecules are:

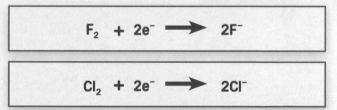

$$F_2 + 2e^- \longrightarrow 2F^-$$

$$Cl_2 + 2e^- \longrightarrow 2Cl^-$$

By looking at an equation of a reaction, you can decide whether it is **oxidation** or **reduction**:
- If **electrons** are **added**, it's a **reduction** reaction.
- If **electrons** are **taken away**, it's an **oxidation** reaction.

So, the reactions shown above are reduction reactions.

An easy way to remember the definitions of oxidation and reduction is by remembering **OILRIG**:
- **O**xidation **I**s **L**oss of electrons.
- **R**eduction **I**s **G**ain of electrons.

C4 Transition Elements

The Transition Metals

The **transition metals**, a block of metallic elements, are between Groups 2 and 3 of the periodic table. This block includes iron (Fe), copper (Cu), platinum (Pt), mercury (Hg), chromium (Cr) and zinc (Zn).

The Transition Metals

Transition metals have the typical properties of metals. Their compounds are often coloured, for example:

- **copper** compounds are **blue**
- **iron(II)** compounds are **light green**
- **iron(III)** compounds are **orange–brown**.

Many transition metals and their compounds are catalysts, for example:

- iron is used in the Haber process
- nickel is used in the manufacture of margarine.

Thermal Decomposition

Thermal decomposition is a reaction where a substance is broken down into two or more substances by heating.

When **transition metal carbonates** are heated, a **colour change** happens. They decompose (break down) to form a **metal oxide** and **carbon dioxide**. The test for carbon dioxide is that it **turns limewater milky**.

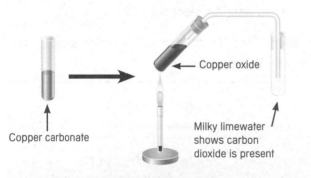

Copper carbonate

Copper oxide

Milky limewater shows carbon dioxide is present

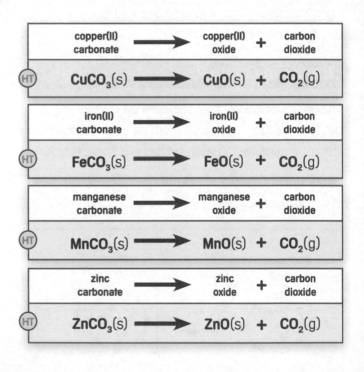

copper(II) carbonate → copper(II) oxide + carbon dioxide

(HT) $CuCO_3(s) \longrightarrow CuO(s) + CO_2(g)$

iron(II) carbonate → iron(II) oxide + carbon dioxide

(HT) $FeCO_3(s) \longrightarrow FeO(s) + CO_2(g)$

manganese carbonate → manganese oxide + carbon dioxide

(HT) $MnCO_3(s) \longrightarrow MnO(s) + CO_2(g)$

zinc carbonate → zinc oxide + carbon dioxide

(HT) $ZnCO_3(s) \longrightarrow ZnO(s) + CO_2(g)$

Identifying Transition Metal Ions

Precipitation is the reaction between **solutions** that makes an **insoluble solid**. The insoluble solid is known as a precipitate.

The following ions form coloured **precipitates**:

Metal Ion	Symbol	Colour	Equation
Copper(II)	Cu^{2+}	Blue	(HT) $Cu^{2+} + 2OH^- \longrightarrow Cu(OH)_2$
Iron(II)	Fe^{2+}	Grey-green	(HT) $Fe^{2+} + 2OH^- \longrightarrow Fe(OH)_2$
Iron(III)	Fe^{3+}	Orange	(HT) $Fe^{3+} + 3OH^- \longrightarrow Fe(OH)_3$

Metal Structure and Properties C4

Metals

Transition metals have many uses, for example:

- **Iron** is used to make steel (which is used to make cars and bridges because it's very strong).
- **Copper** is used to make electrical wiring because it's a good conductor.

Metals are very useful materials because of their properties. Several of their properties include:

- They're **lustrous**, e.g. gold is used in jewellery.

- They're **hard** and have a **high density**, e.g. steel is used to make drill parts.
- They have **high tensile strength** (able to bear loads), e.g. steel is used to make bridge girders.
- They have **high melting** and **boiling points**, e.g. tungsten is used to make light-bulb filaments.
- They're **good conductors** of **heat** and **electricity**, e.g. copper is used to make pans and wiring.

Structure of Metals

Metal atoms are packed very close together in a regular arrangement. The atoms are held together by **metallic bonds**.

Metals have **high melting and boiling points** because lots of energy is needed to break the strong metallic bonds. As the metal atoms pack together, they build a structure of crystals.

HT Metal crystals are made from closely packed positive metal ions in a 'sea' of **delocalised** (free) electrons. The **free movement** of the electrons allows the metal to **conduct electricity**.

The metal is held together by **strong forces** called metallic bonds. These are the electrostatic attractions between the metal ions and the delocalised electrons. So the metals have high melting and boiling points.

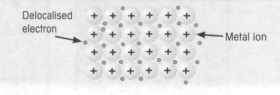

Delocalised electron   Metal ion

Superconductors

Metals are able to conduct electricity because the electrons flow easily through them, moving from atom to atom. At low temperatures, some metals can become **superconductors**.

A superconductor has little, or no, resistance to the flow of electricity. This low resistance is useful for:

- powerful electromagnets, e.g. inside medical scanners
- very fast electronic circuits, e.g. in a supercomputer
- power transmission that doesn't lose energy.

HT The disadvantage of current superconductors is that they only work at temperatures **below −200°C**. This very low temperature is **costly** to maintain and impractical for large-scale uses. So, there is a need to develop superconductors that will work at room temperature (20°C).

Quick Test

1. Why do Group 7 metals have similar chemical properties?
2. Which transition metal is used as a catalyst in the Haber process?
3. What is a thermal decomposition reaction?

C4 Purifying and Testing Water

Water

The four main sources of water in the UK are:
- rivers
- lakes
- reservoirs
- aquifers (wells and bore holes).

Water is an important resource for industry as well as being essential for drinking, washing, etc. The **chemical industry** uses water:
- as a coolant
- as a solvent
- as a raw material.

In some parts of Britain, the **demand** for water is **higher than the supply**, so it's important to **conserve** and not waste water.

Many parts of the developing world don't have access to clean water (water without disease-carrying microorganisms). The World Health Organisation estimates that:
- over 2 million people worldwide die every year from water-borne diseases
- nearly 20% of the world's population doesn't have access to clean drinking water.

Water Treatment

Water Treatment Process

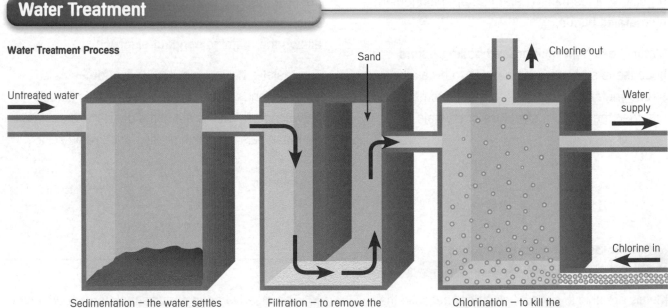

Sand

Chlorine out

Untreated water

Water supply

Chlorine in

Sedimentation – the water settles to allow the insoluble particles to sink

Filtration – to remove the very fine particles

Chlorination – to kill the microorganisms in the water

Water has to be treated to purify it and make it safe to drink. Untreated (raw) water can contain:
- **insoluble** particles
- pollutants
- microorganisms
- dissolved **salts** and minerals.

(HT) Tap water isn't pure – it contains soluble materials that aren't removed by the normal water treatment process. Some of these materials could be poisonous, so extra steps must be taken to remove them.

To obtain pure water, it must be **distilled**, but this process uses a lot of energy and is expensive.

The equipment and energy needed to **distil sea water** is very expensive. The cost of making drinking water out of sea water is currently too high to make it a realistic option in the UK.

Insoluble • Salt • Distillation

Pollutants in Water

Pollutants that can be found in water supplies are often difficult to remove. They include:
- **nitrates** from the run-off of fertilisers
- **lead compounds** from old pipes in the plumbing
- **pesticides** from spraying crops near to the water supply.

Dissolved Ions

The dissolved **ions** of some salts are easy to identify as they will undergo **precipitation** reactions. A precipitation reaction occurs when an insoluble solid is made from mixing two solutions together.

Sulfates can be detected using barium chloride solution – a white precipitate of barium sulfate forms. For example:

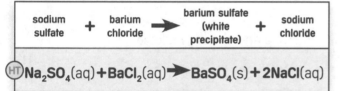

$$\text{(HT) } Na_2SO_4(aq) + BaCl_2(aq) \rightarrow BaSO_4(s) + 2NaCl(aq)$$

Silver nitrate solution is used to detect **halide ions**. Halides are the ions made by the halogens (Group 7).

With silver nitrate:
- chlorides form a white precipitate
- bromides form a cream precipitate
- iodides form a pale yellow precipitate.

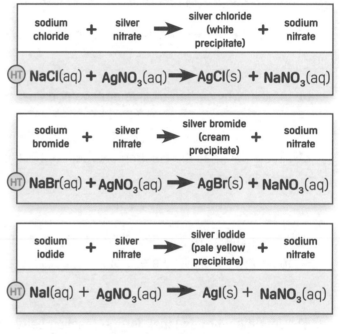

$$\text{(HT) } NaCl(aq) + AgNO_3(aq) \rightarrow AgCl(s) + NaNO_3(aq)$$

$$\text{(HT) } NaBr(aq) + AgNO_3(aq) \rightarrow AgBr(s) + NaNO_3(aq)$$

$$\text{(HT) } NaI(aq) + AgNO_3(aq) \rightarrow AgI(s) + NaNO_3(aq)$$

Interpreting Data

In your exam, you may be asked to interpret data about water resources in the UK. For example, this table shows some pollutants and the maximum amounts allowed in drinking water.

You don't have to remember the data, but you might, for example, be asked to pick out which pollutant has the smallest allowed concentration, or transfer this data onto a graph.

Pollutant	Maximum Amount Allowed
Nitrates	50 parts in 1 000 000 000 parts water
Lead	50 parts in 1 000 000 000 parts water
Pesticides	0.5 parts in 1 000 000 000 parts water

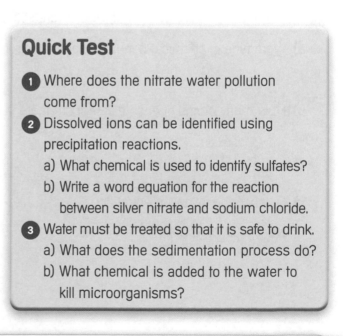

Quick Test

1. Where does the nitrate water pollution come from?
2. Dissolved ions can be identified using precipitation reactions.
   a) What chemical is used to identify sulfates?
   b) Write a word equation for the reaction between silver nitrate and sodium chloride.
3. Water must be treated so that it is safe to drink.
   a) What does the sedimentation process do?
   b) What chemical is added to the water to kill microorganisms?

C4 Exam Practice Questions

1 Sodium metal can react with chlorine gas to make sodium chloride.

a) What is an ion? **[1]**

b) How would you make a 1^- ion from a neutral atom? **[1]**

c) Give the properties of a typical ionic substance such as sodium chloride. **[3]**

2 There are about 100 naturally occurring elements. They are all listed in the periodic table.

a) What type of elements usually bond covalently? **[1]**

b) Explain why carbon dioxide and water do not conduct electricity. **[1]**

c) An element has the electronic structure 2.8.4. In which group and period would you find this element? **[1]**

d) Alkali metals are found in Group 1 of the periodic table. These metals all have 1 electron in their outer shell. Why are Group 1 metals stored in oil? **[2]**

e) Write the word equation for the reaction between lithium and water. **[2]**

3 Atoms often form bonds so that they have a complete outer shell of electrons.

a) What type of chemical bonding shares electron pairs? **[1]**

b) How many atoms of each element are there in $Mg(NO_3)_2$? **[1]**

c) Balance the following equation:

$Na + H_2O \rightarrow NaOH + H_2$ **[1]**

4 When some metal carbonates are heated they undergo a chemical change.

a) What is thermal decomposition? **[2]**

b) Write the word equation for the thermal decomposition of copper carbonate. **[2]**

c) Which metal ion makes an orange–brown precipitate with sodium hydroxide solution? **[1]**

5 Halogens are the elements found in Group 7 of the periodic table.

a) Which gas is used to sterilise water? **[1]**

b) Write the word equation to show the reaction of sodium iodide solution with chlorine gas. **[2]**

HT

c) Write the balanced symbol equation to show the reaction of sodium iodide solution with chlorine gas. **[2]**

d) Write the ionic equation to show the formation of bromide ions from a bromine molecule. **[2]**

6 Halogens are found in Group 7 of the periodic table. Explain why halogens have similar properties and describe the trend in reactivity as you go down the group. **[6]**

✎ The quality of your written communication will be assessed in your answer to this question.

C5 Moles and Molar Mass

Molar Mass

The amount of substance in a chemical reaction is measured in **moles**. The mass of one mole (the **molar mass**) of any substance is the **relative formula mass (M_r)** in grams (g). Molar mass is measured in g/mol.

Example

What is the molar mass of magnesium hydroxide, $Mg(OH)_2$?

Mg $1 \times 24 = 24$
O $2 \times 16 = 32$
H $2 \times 1 = 2$
M_r $24 + 32 + 2 = 58g/mol$

The M_r of $Mg(OH)_2$ is 58, so the mass of 1 mole of $Mg(OH)_2$ is **58g**.

HT Molar Mass, Moles and Mass

The relative atomic mass of an element is the average mass of the atoms of that element compared with a twelfth of the mass of a carbon-12 (^{12}C) atom.

You can use this formula to calculate the number of moles of an element or a compound:

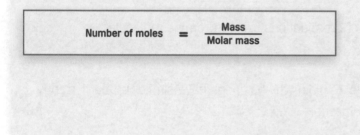

$$\text{Number of moles} = \frac{\text{Mass}}{\text{Molar mass}}$$

Example 1

How many moles of ethanol are there in 230g of ethanol? (The M_r of ethanol is 46.)

$$\text{Number of moles} = \frac{\text{Mass}}{\text{Molar mass}}$$

$$= \frac{230g}{46g} = 5 \text{ moles}$$

Example 2

What is the mass of oxygen in 3 moles of aluminium oxide (Al_2O_3)?

$$\text{Mass} = \text{Number of moles} \times \text{Molar mass}$$

$$= 9 \times 16 = \textbf{144g}$$

3 moles of aluminium oxide contains 9 (3×3) moles of oxygen.

Conservation of Mass

During a chemical reaction, no mass is lost or gained, i.e. it's **conserved**. But, the mass measured at the end of a reaction might be:

* greater if a gas has been gained from the air **or**
* less if water vapour or a gas has been allowed to escape.

The mass of gas made or lost can be determined by calculating the mass change.

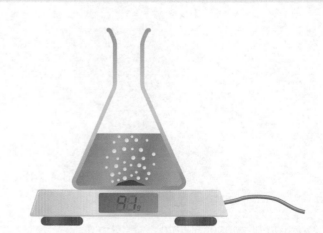

Mass drops as the gas is made in the chemical reaction and lost to the atmosphere.

Examples of Conservation

Example 1

When 50g of calcium carbonate is heated in a thermal decomposition reaction, 28g of calcium oxide is made. What mass of carbon dioxide is lost?

calcium carbonate	➡	calcium oxide	+	calcium dioxide

(HT)

$CaCO_3(s)$	➡	$CaO(s)$	+	$CO_2(g)$

50g = 28g + Mass of carbon dioxide
Mass of carbon dioxide = 50g − 28g = **22g**

(HT)

$CaCO_3$	➡	CaO	+	CO_2

$$\frac{50}{100} \qquad \frac{28}{56} \qquad \frac{22}{44}$$

= 0.5 mole   = 0.5 mole   = **0.5 mole**

So, 0.5 mole of carbon dioxide is made.
44 × 0.5 = 22g of carbon dioxide is made.

Example 2

When 12g of carbon is burned in oxygen it makes 44g of carbon dioxide. What mass of oxygen has reacted?

carbon	+	oxygen	➡	calcium dioxide

12g   +   Mass of oxygen   ➡   44g

Mass of oxygen = 44g − 12g = **32g**

C	+	O_2	➡	CO_2

$$\frac{12}{12} \qquad \frac{32}{32} \qquad \frac{44}{44}$$

= 1 mole   = **1 mole**   = 1 mole

So, 1 mole of oxygen molecules is needed with a mass of 32g.

Reacting Ratios

If you know the reacting masses in a reaction, you can calculate further reacting masses using **ratios**.

Example

The reaction between 160g of copper sulfate and 106g of sodium carbonate produces 124g of copper carbonate and 142g of sodium sulfate. How much copper sulfate and sodium carbonate are needed to produce 372g of copper carbonate?

$CuSO_4(aq) + Na_2CO_3(aq)$ ➡ $CuCO_3(s) + Na_2SO_4(aq)$

160g + 106g = 124g + 142g

372g ÷ 3 = 124g, so you just need to multiply all the above masses by the same amount (×3) to find the new set of masses.

(3 × 160g) + (3 × 106g) = (3 × 124g) + (3 × 142g)
= 480g + 318g = 372g + 426g

So, **480g** of copper sulfate and **318g** of sodium carbonate are needed.

N.B. A quick way to check that your calculation is correct is to add the masses of the reactants together to check that the total mass of the reactants is equal to the total mass of the products.

Quick Test

1. What is a mole?
2. What is the molar mass (M_r) of carbon dioxide (CO_2)?
3. Hydrogen will burn in oxygen to make water.
   a) When 4g of hydrogen is reacted with 32g of oxygen, what mass of water is made?
   b) What is the molar mass (M_r) of water?
   (HT) c) How many moles of water are made when 8g of hydrogen is reacted with excess oxygen?

Mass of Elements in a Compound

The **mass** of a compound is made up of the masses of all its elements added together. So, if you know the mass of a compound and the mass of one of the elements, you can calculate the mass of the other element.

Example

80g of copper oxide contains 16g of oxygen. What mass of copper does it contain?

Mass of copper + 16g = 80g
Mass of copper = 80g − 16g = **64g**

Empirical Formula

The **empirical formula** is the simplest whole number ratio of each type of atom in a compound. For example, all alkenes have the empirical formula C_1H_2, or CH_2.

You can work out the empirical formula of a substance from its chemical formula, for example, the empirical formula of ethanoic acid (CH_3COOH) is CH_2O.

HT The empirical formula of a compound can be calculated from **either**:

- the percentage composition of the compound by mass **or**
- the mass of each element in the compound.

To calculate the empirical formula:

1 List all the elements in the compound.
2 Divide the data for each element by its A_r (to find out the number of moles).
3 Select the smallest answer from step 2 and divide each answer by that result to obtain a ratio.
4 The ratio may have to be scaled up to give whole numbers.

Example 1

What is the empirical formula of a hydrocarbon containing 75% carbon. (Hydrogen = 25%)

HT You can also calculate the percentage of an element in a compound.

Example

Calculate the percentage of nitrogen in ammonia (NH_3) where A_r for N = 14 and A_r for H = 1.

$$\% \text{ nitrogen} = \frac{14}{14 + (3 \times 1)} \times 100 = \textbf{82\%}$$

Quick Test

1 What is the empirical formula for benzene (C_6H_6)?
2 Ammonium chloride (NH_4Cl) can be used as a fertiliser. What is the molar mass (M_r) of this fertiliser?
3 HT A hydrocarbon contains 24g of carbon and 4g of hydrogen. What is the empirical formula of this compound?

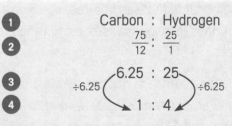

So, the empirical formula is C_1H_4, or **CH_4**

Example 2

What is the empirical formula of a compound containing 24g of carbon, 8g of hydrogen and 32g of oxygen?

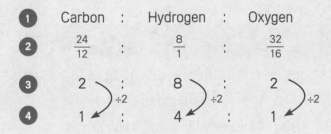

So, the empirical formula is CH_4O.

Volume

The two units commonly used to measure the **volume** of liquids and solutions are:

- cm^3 (cubic centimetres)
- dm^3 (cubic decimetres). $1dm^3$ is equal to $1000cm^3$ and is known as 1 litre.

To convert a volume:

- from cm^3 into dm^3, divide it by 1000
- from dm^3 into cm^3, multiply it by 1000.

Example

Convert $2570cm^3$ into dm^3.

$$\text{Volume in } dm^3 = \frac{\text{Volume in } cm^3}{1000}$$

$$= \frac{2570}{1000} = \mathbf{2.57dm^3}$$

$1cm^3$

1cm
1cm
1cm

$1000cm^3$ $(1dm^3)$

10cm
10cm
10cm
$1dm^3$

Concentration

The **concentration** of a solution can be measured in:

- g/dm^3 (grams per cubic decimetre)
- mol/dm^3 (moles per cubic decimetre).

In a concentrated solution, the solute particles are more crowded together than they are in a dilute solution.

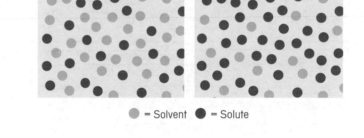

Dilute   Concentrated

● = Solvent   ● = Solute

Dilution

A concentrated solution can be made more **dilute** by adding water.

Some concentrated solutions must be diluted before they're used, for example, orange cordial has to be diluted before you can drink it so that it doesn't taste too strong.

It's important to accurately follow the dilution instructions. If a medicine is too dilute then it will not work properly, and if it's too concentrated it may even make you more ill. If baby milk is not diluted correctly then it could harm the baby.

Example

$5cm^3$ of a solution has a concentration of $1mol/dm^3$. How much water should be added to the solution in order to make it have a concentration of $0.1mol/dm^3$?

> Since $0.1mol/dm^3$ is ten times smaller than $1mol/dm^3$, the volume of the solution should be ten times greater.

Volume of water added $= 9 \times 5cm^3$

$$= \mathbf{45cm^3}$$

C5 Quantitative Analysis

Guideline Daily Amounts

The **guideline daily amount** (**GDA**) informs you how much of a nutrient a person needs each day for a healthy diet.

You may be asked to read the GDA from a food label.

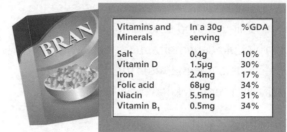

Vitamins and Minerals	In a 30g serving	%GDA
Salt	0.4g	10%
Vitamin D	1.5µg	30%
Iron	2.4mg	17%
Folic acid	68µg	34%
Niacin	5.5mg	31%
Vitamin B_1	0.5mg	34%

HT You may be asked to use a food label to convert the amount of sodium from sodium salt. But the conversion could be inaccurate as sodium ions come from a number of sources.

Example
What mass of sodium ions is in a 30g serving of bran?

1 0.4g of sodium chloride (NaCl).

2 Number of moles of NaCl = $\dfrac{0.4}{58.5}$ = 0.0068 moles.

3 There would be 0.0068 moles of sodium ions.

Mass of sodium ions = 0.0068 × 23 = **0.16g**

Concentration in g/dm³ and mol/dm³

The **concentration** of a solution in g/dm^3 can be calculated using the following formula:

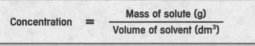

$$\text{Concentration} = \frac{\text{Mass of solute (g)}}{\text{Volume of solvent (dm}^3\text{)}}$$

The concentration of a solution in mol/dm^3 can be calculated using the following formula:

$$\text{Concentration} = \frac{\text{Amount of solute (mol)}}{\text{Volume of solvent (dm}^3\text{)}}$$

Example
2g of sodium chloride is dissolved in $100cm^3$ of water. What is the concentration (in g/dm^3)?

First, change the volume to dm^3... $\dfrac{100cm^3}{1000}$ = $0.1dm^3$

$\text{Concentration} = \dfrac{\text{Mass of solute}}{\text{Volume of solvent}}$

$= \dfrac{2}{0.1} = \textbf{20g/dm}^3$

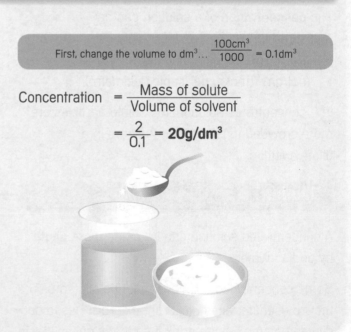

Quick Test

1 Convert $1200cm^3$ into dm^3.

2 What are the two common units of concentration?

3 Explain how you dilute a solution.

4 Why is it important to ensure that medicines are diluted correctly?

Key Words   GDA

Titration

When an acid and an alkali react together, it's known as a **neutralisation** reaction:

| acid | + | alkali | → | salt | + | water |

You can carry out a **titration** to find out how much acid is needed to neutralise an alkali using this method:

1. Measure the alkali into a conical flask by using a pipette and filler. It is important to use the filler to prevent the acid getting into contact with your skin.
2. Add a few drops of **indicator** to the conical flask.
3. Fill the burette with acid.
4. Record the start volume of acid.
5. Add acid slowly to the alkali until the indicator just changes colour (the end point).
6. Record the end volume of acid.
7. Work out how much acid has been added (final volume – start volume). This is called the **titre**. The titre depends on the concentration of the reactants.

8. To improve the accuracy of the results, repeat the titration until you have consistent (concordant) results, then take an average (mean).

N.B. Sometimes you can put the alkali in the burette, and the acid in the conical flask.

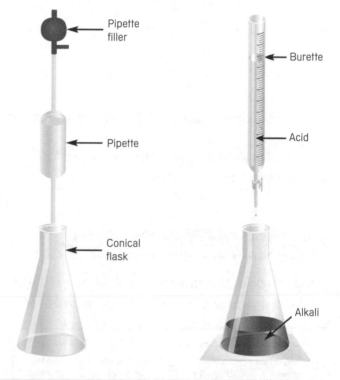

Indicators

Indicators change colour depending on whether they are in acidic or alkaline solutions.

Single indicators, such as litmus, produce a sudden colour change at the end of the titration. This clearly shows the end point.

Universal indicator is a mixture of different indicators, which gives a continuous range of colours. You can estimate the pH of a solution by comparing the colour of the indicator in solution to the pH colour chart.

This table shows the colours of certain indicators in acidic and alkaline solutions:

Indicator	Colour in Acid	Colour in Alkali
Litmus	Red	Blue
Phenolphthalein	Colourless	Pink
Universal indicator	Red	Blue

HT In an acid–base titration, the pH changes very suddenly near the end point, so a single indicator, e.g. litmus, shows the change very clearly.

When a mixed indicator, e.g. universal indicator, is used it's harder to see the end point because it gives a range of colours.

C5 Titrations

pH Curves

pH curves can be drawn to show what happens to the pH in a neutralisation reaction:

- An acid has a low pH. When an alkali is added to it, the pH increases.
- An alkali has a high pH. When an acid is added to it, the pH decreases.

You should be able to read and interpret pH curves (like the one opposite) to work out:

- the titre (the volume of acid needed to neutralise the alkali)
- the pH when a certain amount of acid has been added.

HT You should be able to sketch pH curves for the titration of an acid or an alkali.

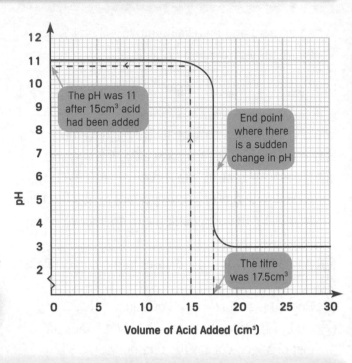

The pH was 11 after 15cm³ acid had been added

End point where there is a sudden change in pH

The titre was 17.5cm³

Volume of Acid Added (cm³)

HT Concentration Formulae

At the end point of a titration where the acid and alkali react in a one-to-one ratio, the number of moles of acid is equal to the number of moles of alkali:

| Concentration of acid | × | Volume of acid | = | Concentration of alkali | × | Volume of alkali |

You can use these formulae to calculate concentration, volume and moles:

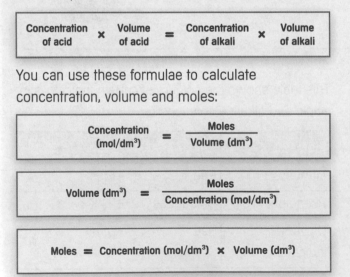

$$\text{Concentration (mol/dm}^3) = \frac{\text{Moles}}{\text{Volume (dm}^3)}$$

$$\text{Volume (dm}^3) = \frac{\text{Moles}}{\text{Concentration (mol/dm}^3)}$$

$$\text{Moles} = \text{Concentration (mol/dm}^3) \times \text{Volume (dm}^3)$$

Example 1

$0.025dm^3$ of a sample of sodium hydroxide (NaOH) is completely neutralised by $0.030dm^3$ of a $0.1mol/dm^3$ hydrochloric acid (HCl). What is the concentration of alkali?

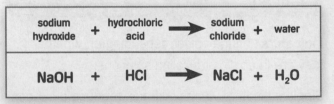

| sodium hydroxide | + | hydrochloric acid | → | sodium chloride | + | water |
| $NaOH$ | + | HCl | → | $NaCl$ | + | H_2O |

Concentration of acid × Volume of acid = Concentration of alkali × Volume of alkali

$0.1 \times 0.030 = $ Concentration of alkali $\times 0.025$

$$\text{Concentration of alkali} = \frac{0.1 \times 0.030}{0.025}$$

$$= 0.12mol/dm^3$$

Quick Test

1. What is an indicator?
2. Describe the changes in pH as an acid is added to an alkali.

Measuring Gas Volumes

You can use the following apparatus to collect and measure the **volume** of a gas made in a reaction:

- An upturned measuring cylinder.
- A gas syringe.
- An upturned burette.

You should use this method to measure the volume of gas produced by an experiment:

1. Measure out the reactants.
2. Add the reactants together in a conical flask and start the stopwatch.
3. Record the volume of gas produced at regular time intervals until the volume stops increasing.

A reaction stops when one of the reactants has been used up. The reactant that gets used up first is called the **limiting reactant**.

In a reaction where there is a one-to-one ratio between reactants, the limiting reactant is the one with the smallest number of moles.

The more reactant that's used, the greater the amount of product (in this case, gas) that's produced.

HT More reactants means more reactant particles and so there will be a greater number of collisions, which increase the number of product particles. The number of particles of the limiting reactant determines the maximum number of product particles that can be made.

The amount of gas produced in a reaction is directly proportional to the amount of limiting reactant used.

Apparatus for Collecting and Measuring Volume of Gas made in a Reaction

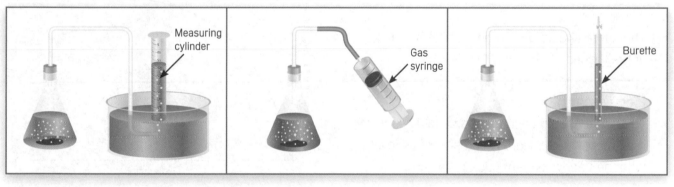

Measuring cylinder

Gas syringe

Burette

Measuring Gas Masses

The amount of gas made in a reaction can be measured by monitoring the change in mass of a reaction. You should use this method to measure the mass of gas produced by an experiment:

1. Measure the mass of an empty conical flask.
2. Measure out the reactants.
3. Record the total mass of the reactants and the flask.
4. Add the reactants together in the flask and start the stopwatch.
5. Record the mass of the flask and reactants at regular time intervals, until the mass stops changing.

Cotton wool

Conical flask

Balance

C5 Gas Volumes

Interpreting Graphs of Reactions

Graphs can be used to show the results of a reaction. You can find out the following information from a graph:

1. The total **volume** of gas produced.
2. When the reaction ended.
3. The volume of gas produced at a particular time (or the time at which a particular volume of gas was produced).
4. The point at which the reaction was fastest.

You may be asked to compare the rate of reaction for different reactions by using the gradient.

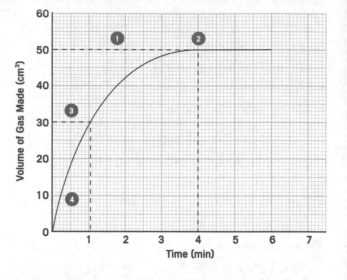

HT Sketching Graphs of Reactions

When you're sketching a graph to show the volume of gas made in a reaction, you should remember the following rules:

- The curve should be steepest at the beginning of the reaction (when the rate is fastest).
- The curve should get shallower as the reaction progresses.
- The curve should become horizontal to show the end of the reaction, which should be level with the final volume of gas produced.

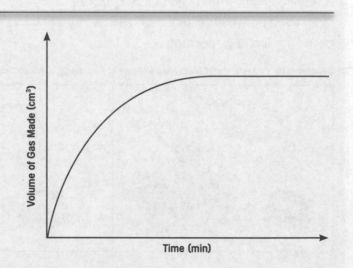

Calculating Volumes and Amounts of Gases

One mole of any gas occupies a volume of **24dm³** at room temperature and pressure.

You can use this rule to:

- calculate the volume of a known amount of gas
- calculate the amount of gas if the volume is known.

Example 1

What is the volume of half a mole of nitrogen?

Volume = 0.5 × 24

= **12dm³**

Example 2

A balloon is filled with oxygen until it has a volume of 6dm³. How many moles of oxygen are in the balloon?

Moles = $\frac{6}{24}$

= **0.25 mole**

Reversible Reactions

A **reversible reaction** can go forwards or backwards, under the same conditions. It's represented by ⇌. For example, the reaction between nitrogen and hydrogen to produce ammonia is reversible.

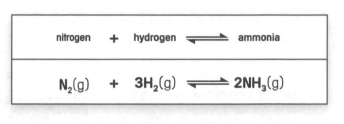

| nitrogen | + | hydrogen | ⇌ | ammonia |

$$N_2(g) \ + \ 3H_2(g) \ \rightleftharpoons \ 2NH_3(g)$$

Equilibrium

A reversible reaction can reach **equilibrium** (a balance). This means that the rate of the forward reaction is equal to the rate of the backward reaction. At equilibrium, the amounts and concentrations of reactants and products stay the same, even though reactions are still taking place.

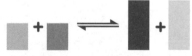

| A | + | B | ⇌ | C | + | D |
Reactants   Products

The position of the equilibrium can be altered by changing:

- the temperature
- the pressure
- the concentration of reactant(s) and/or product(s).

If the position of the equilibrium lies to the right of the reaction equation, the concentration of the products is greater than the concentration of the reactants.

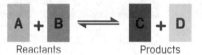

If the position of the equilibrium lies to the left of the reaction equation, the concentration of the products is less than the concentration of the reactants.

You should be able to read tables or graphs of equilibrium composition to obtain the following information:

- The composition at a particular temperature.
- The composition at a particular pressure.
- The effect of temperature and pressure on composition.

Example

The table and graphs show how altering the reaction conditions can change the equilibrium composition for the reaction to make ammonia.

We can see that percentage of ammonia made falls when the temperature increases.

Pressure (atmospheres)	Ammonia made at 300°C (%)	Ammonia made at 600°C (%)
100	43	4
200	62	12
300	74	18
400	79	19
500	80	20

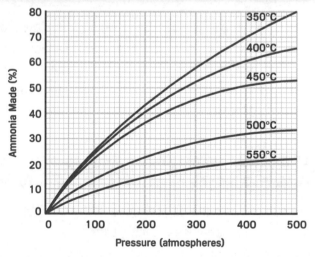

From the table we can see that 200 atmospheres and 600°C is 12%.

From the graph we can see that the percentage of ammonia made at 300 atmospheres and 400°C is 52%.

C5 Equilibria

The Contact Process

The raw materials sulfur, air and water are made into sulfuric acid in the **Contact Process**:

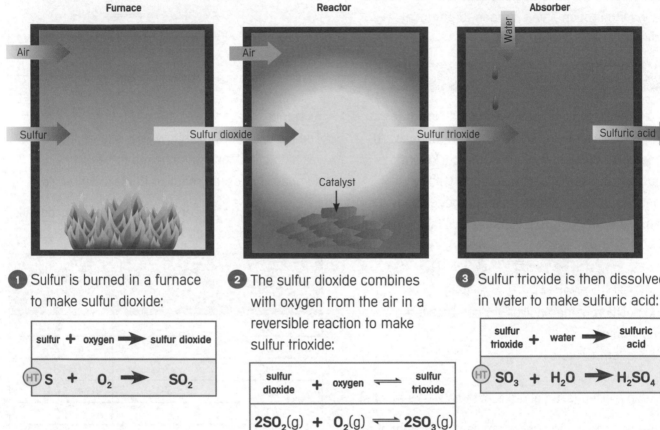

| Furnace | Reactor | Absorber |

1 Sulfur is burned in a furnace to make sulfur dioxide:

sulfur + oxygen ➡ sulfur dioxide
(HT) S + O_2 ➡ SO_2

2 The sulfur dioxide combines with oxygen from the air in a reversible reaction to make sulfur trioxide:

sulfur dioxide + oxygen ⇌ sulfur trioxide
$2SO_2(g) + O_2(g) \rightleftharpoons 2SO_3(g)$

This reaction takes place using a vanadium(V) oxide (V_2O_5) **catalyst**, at a temperature of about 450°C and at atmospheric pressure.

3 Sulfur trioxide is then dissolved in water to make sulfuric acid:

sulfur trioxide + water ➡ sulfuric acid
(HT) $SO_3 + H_2O$ ➡ H_2SO_4

Uses of Sulfuric Acid

The sulfuric acid that is produced in the Contact Process has many uses, such as the manufacturing of:
- paints and pigments
- soaps and detergents
- fibres
- plastics
- fertilisers.

HT Changing Equilibrium Conditions

A reversible reaction will only reach equilibrium if the conditions (such as temperature and pressure) aren't changed and no substance is added or removed. This is known as a **closed system**.

At the start of an equilibrium reaction, the forward reaction will slow down and the backward reaction will speed up until both reactions are at the same rate.

The equilibrium can be moved to the right of the reaction equation by:

* adding more reactant **or**
* removing the product as it's made.

The equilibrium can be moved to the left of the reaction equation by:

* reducing the amount of reactant **or**
* increasing the amount of product.

In a reaction that involves gases, an increase in pressure moves the equilibrium in the direction that has the fewest moles of gas.

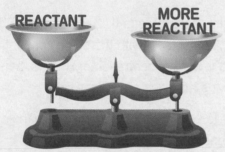

Conditions in the Contact Process

A catalyst is used in the reaction to speed up the rate of production of sulfur trioxide but it doesn't change the position of the equilibrium.

Increasing the temperature increases the rate of the reaction but it also reduces the yield and pushes the equilibrium position to the left.

A **compromise** temperature of about 450°C is used to get a balance between yield and rate.

A higher pressure would push the equilibrium to the right and increase the yield. But the extra cost of increasing the pressure isn't worth the small amount of increase in yield because the equilibrium position is well over to the right.

Quick Test

1. What two variables can be used to monitor the amount of gas made in a chemical reaction?
2. What is the symbol used to show a reversible reaction?
3. What four factors can be altered in order to change the position of equilibrium?
4. The Contact Process is used in industry to make sulfuric acid. In the reactor sulfur dioxide is converted into sulfur trioxide. Write a balanced symbol equation for this reaction.

C5 Strong and Weak Acids

Strong and Weak Acids

Acids **ionise** in water to make hydrogen ions (H^+):

- A **strong acid** ionises completely in water.
- A **weak acid** only partially ionises in water. The ionisation of a weak acid is a reversible reaction, so it makes an equilibrium mixture.

Strong acids, e.g. hydrochloric acid, nitric acid and sulfuric acid, have a lower pH than weak acids, e.g. ethanoic acid, if they're of the same concentration.

Ethanoic acid or hydrochloric acid react with:

- magnesium to produce hydrogen gas
- calcium carbonate to produce carbon dioxide gas.

If an equal amount of ethanoic acid and hydrochloric acid are used in these reactions, the same volumes of gas will be made. The volume of gas made is determined by the amount of reactants used, not by the acid's strength. But, the reaction with ethanoic acid is slower as there are fewer hydrogen ions than in the same concentration of hydrochloric acid, and so there are fewer collisions.

HT The **concentration** of an acid is determined by how many moles of the acid are dissolved in $1dm^3$.

The **strength** of an acid is determined by how much it ionises. A strong acid produces more H^+ ions than a weak acid of the same concentration, because the weak acid does not ionise completely.

- Hydrochloric acid completely ionises:

$$HCl \longrightarrow H^+ + Cl^-$$

- Ethanoic acid partially ionises:

$$CH_3COOH \rightleftharpoons CH_3COO^- + H^+$$

An acid with more H^+ ions (i.e. a strong acid) has a lower pH. A weak acid with the same concentration will have a lower concentration of H^+ ions than a diluted strong acid and so a higher pH.

Hydrochloric acid reacts quicker than ethanoic acid because:

- hydrochloric acid is a stronger acid than ethanoic acid
- hydrochloric acid has a greater concentration of hydrogen ions than ethanoic acid
- the greater concentration of hydrogen ions in hydrochloric acid leads to a higher frequency of collision between hydrogen ions and the other reactant.

Electrolysis of Acids

Acids conduct electricity. A strong acid, such as hydrochloric acid, is a better conductor than the same concentration of a weak acid, such as ethanoic acid. This is because there are fewer hydrogen ions in ethanoic acid to carry the charge.

When hydrochloric acid or ethanoic acid is used in electrolysis, hydrogen gas is made at the negative electrode.

This is because when electricity is passed through the acid, the hydrogen ions are attracted to the negative electrode, where they become hydrogen molecules.

HT The greater the concentration of hydrogen ions in an acid, the greater the electrical conductivity as the ions carry the charge. This is why strong acids, for example, hydrochloric acid, are better conductors than weak acids.

Key Words   **Ionise • Strong acid • Weak acid**

Precipitation Reactions

In a solid ionic substance, the ions are in fixed positions, but when they dissolve in water they are free to move about.

A **precipitation** reaction occurs when a **precipitate** (an insoluble solid) is made by mixing two ionic solutions together. The precipitate (the product) is made when ions from one solution collide with ions from the other solution and form an insoluble compound.

You can use this method to make an insoluble compound:

1 Mix the reactant solutions.

2 Filter off the precipitate.

3 Wash the residue in the filter funnel with a little distilled water.

4 Dry the residue (the product) in an oven at 50°C.

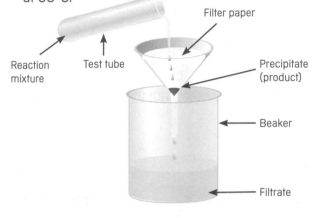

Detecting Ions

Halides are the ions made by the halogens (Group 7 elements). You can use silver nitrate solution to detect halide ions, e.g. with silver nitrate:

- chlorides (Cl^-) form a white precipitate
- bromides (Br^-) form a cream precipitate
- iodides (I^-) form a yellow precipitate.

You can detect sulfate ions by using barium chloride solution. It will cause a white precipitate of barium sulfate to form:

Precipitation reactions are very fast reactions between ions. For example, when sodium chloride and silver nitrate react, the precipitate of silver chloride forms almost instantly:

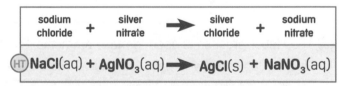

HT $NaCl(aq) + AgNO_3(aq) \longrightarrow AgCl(s) + NaNO_3(aq)$

HT The ions involved in this reaction are Na^+, Cl^-, Ag^+ and NO_3^-. You can write an ionic equation by picking out the ions that react to form the precipitate, for example:

$$Ag^+(aq) + Cl^-(aq) \longrightarrow AgCl(s)$$

The Na^+ and the NO_3^- ions stay dissolved in the water and don't do anything. So, they are called **spectator ions**.

Quick Test

1 What is a strong acid?

2 Explain how a precipitate is formed.

3 What chemical is used to detect sulfate ions?

4 HT Explain why $0.1mol/dm^3$ of nitric acid would have a lower pH than $0.1mol/dm^3$ of ethanoic acid.

C5 Exam Practice Questions

1 Volumes can be measured using different units.

Convert the volume 150cm³ into dm³. **[1]**

2 Titrations can be used to work out the concentration of a solution. What is a titre? **[1]**

3 In a neutralisation reaction acids and bases react together.

Describe how the pH changes when acid is added to an alkali until there is more than enough acid to completely react with the alkali. Tick (✓) the correct answer.

It starts at 7 and falls lower. ⬭

It starts at 7 and rises higher. ⬭

It starts higher than 7 and falls lower. ⬭

It starts lower than 7 and rises higher. ⬭

[1]

4 Methane is used as a hydrocarbon fuel in most UK houses for cooking and heating.

What is the volume of two moles of methane gas at room temperature and pressure? **[2]**

5 The Haber process is used to make ammonia. By changing the conditions, the yield can be altered. Use the table below to help you answer the questions that follow.

Pressure (atmospheres)	Ammonia made at 300°C (%)	Ammonia made at 600°C (%)
100	43	4
200	62	12
300	74	18
400	79	19
500	80	20

a) What happens to the yield as the pressure is increased? **[1]**

b) What happens to the yield as temperature is increased? **[1]**

c) What is the yield at 300°C and 400 atmospheres pressure? **[1]**

6 The Contact Process is used to make sulfuric acid.

a) List the raw materials needed for the Contact Process. **[2]**

b) A lower temperature gives a higher yield in the Contact Process.

Explain why a relatively high temperature of 450°C is used instead. **[1]**

7 When Group 7 elements react they often become halide ions.

Which type of halide gives a cream precipitate when tested with silver nitrate solution? **[1]**

HT

8 Sodium hydroxide has a pH greater than 7 and is an alkali.

a) How many moles are there in 16g of sodium hydroxide? **[2]**

b) What mass of oxygen (A_r = 16) is in 3 moles of sodium hydroxide? **[1]**

9 A titration is carried out in which the acid and alkali react in a 1 : 1 ratio.

$25cm^3$ of an unknown concentration of alkali was completely reacted with $22.5cm^3$ of a $0.2mol/dm^3$ acid.

What was the concentration of the alkali? Ring the correct answer.

      $0.018mol/dm^3$ **$3.6mol/dm^3$**   **$0.18mol/dm^3$**   **$0.36mol/dm^3$**

[1]

10 Hydrochloric acid is a strong acid with a low pH.

Write a balanced symbol equation to show the ionisation of hydrochloric acid in water. **[3]**

11 Precipitation reactions can be used to determine which halide ion is present.

Write a balanced symbol equation for the reaction of potassium iodide (KI) with silver nitrate ($AgNO_3$). **[2]**

C6 Electrolysis

Electrolysis

The **ions** in:
- an **ionic solid** are fixed and can't move
- an **ionic substance** that is **molten** or in **solution** are free to move.

Electrolysis is a chemical reaction in which an ionic liquid is broken down (**decomposed**) into its elements using an **electric current**. It is a flow of charge produced by moving ions and ions are discharged at the electrodes.

The ionic substance is called the **electrolyte**. It must be molten or in solution as the ions need to be free to move:
- The positive ions (**cations**) move to, and discharge at, the negative electrode (**cathode**).
- The negative ions (**anions**) move to, and discharge at, the positive electrode (**anode**).

Electrons are removed from negative ions.

The electrons then flow around the circuit to the negative electrode and are passed to the positive ions.

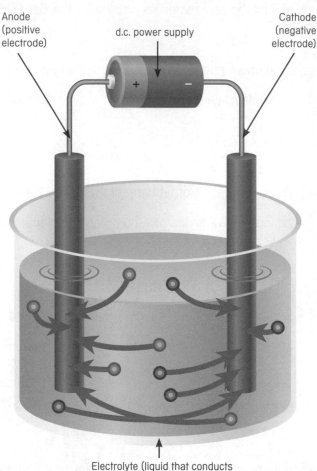

Anode (positive electrode)

d.c. power supply

Cathode (negative electrode)

Electrolyte (liquid that conducts and decomposes in electrolysis)

Amount of Substance Produced

The amount of substance made in electrolysis is determined by the size of the current and the length of time it flows for.

HT The quantity of electricity (Q) passed in an electrolysis reaction can be calculated using the following formula:

Quantity of electricity (coulombs)	=	Current (amps)	×	Time (seconds)
Q	=	I	×	t

One mole of a substance with a 1⁺ charge will be deposited by 96 500 coulombs. This can be used to calculate how much of a substance will be made from the current, and the time the electricity has been on.

More substance is made if:
- a larger current flows
- the current flows for a longer time.

Example

How many moles of silver are deposited when a solution containing Ag^+ ions is electrolysed for 24 125 seconds by a current of 2 amps?

$Q = I \times t$

$= 2 \times 24\ 125 = 48\ 250$ coulombs

Number of moles $= \dfrac{48\ 250}{96\ 500}$

$= 0.5$

Therefore, **0.5 mole Ag^+** is deposited.

Ion • Electrolysis • Electrolyte • Cathode • Anode

Electrolysis of Copper(II) Sulfate

When an electric current is passed through the copper(II) sulfate solution using copper electrodes:

- the positive electrode bubbles as oxygen is made, and the mass decreases because ions move **from** it
- the negative electrode becomes plated with copper and the mass increases because ions move **to** it
- the electrolyte will become less blue as copper(II) sulfate decomposes.

(HT) In the electrolysis of $CuSO_4(aq)$ with copper electrodes:

- the following reaction takes place at the **cathode**:

$$Cu^{2+} + 2e^- \rightarrow Cu$$

- the following reaction takes place at the **anode**:

$$4OH^- - 4e^- \rightarrow O_2 + 2H_2O$$

The amount of substance produced is directly proportional to the time and the current.

Products of Electrolysis

The table below shows the elements made when certain electrolytes undergo electrolysis.

(HT) When **aqueous solutions** undergo electrolysis, it's often easier to decompose the water than to decompose the compound that is dissolved in it. This is the reason why hydrogen and oxygen are produced in the electrolysis of electrolytes such as $NaOH(aq)$ and $K_2SO_4(aq)$.

Quick Test

1. Which electrode are positive ions attracted to?
2. What two factors affect the amount of substance made during electrolysis?

Test for hydrogen gas: a lighted splint causes a 'pop!'.
Test for oxygen gas: a glowing splint is re-lit.

Liquid	State	Elements Made	At the Cathode	At the Anode
Aluminium oxide $Al_2O_3(l)$	Liquid	Aluminium, oxygen	(HT) $Al^{3+}(l) + 3e^- \rightarrow Al(l)$	$2O^{2-}(l) - 4e^- \rightarrow O_2(g)$
Lead bromide $PbBr_2(l)$	Liquid	Lead, bromine	(HT) $Pb^{2+}(l) + 2e^- \rightarrow Pb(l)$	$2Br^-(l) - 2e^- \rightarrow Br_2(g)$
Lead iodide $PbI_2(l)$	Liquid	Lead, iodine	(HT) $Pb^{2+}(l) + 2e^- \rightarrow Pb(l)$	$2I^-(l) - 2e^- \rightarrow I_2(g)$
Potassium chloride $KCl(l)$	Liquid	Potassium, chlorine	(HT) $K^+(l) + e^- \rightarrow K(l)$	$2Cl^-(l) - 2e^- \rightarrow Cl_2(g)$
Sodium hydroxide $NaOH(aq)$	Dissolved in water	Hydrogen, oxygen	(HT) $2H^+(aq) + 2e^- \rightarrow H_2(g)$	$4OH^- - 4e^- \rightarrow O_2 + 2H_2O$
Potassium nitrate $KNO_3(aq)$	Dissolved in water	Hydrogen, oxygen	(HT) $2H^+(aq) + 2e^- \rightarrow H_2(g)$	$4OH^- - 4e^- \rightarrow O_2 + 2H_2O$
Sulfuric acid $H_2SO_4(aq)$	Dissolved in water	Hydrogen, oxygen	(HT) $2H^+(aq) + 2e^- \rightarrow H_2(g)$	$4OH^- - 4e^- \rightarrow O_2 + 2H_2O$

C6 Energy Transfers – Fuel Cells

Reacting Oxygen with Hydrogen

The reaction between hydrogen and oxygen releases energy and is an **exothermic** reaction.

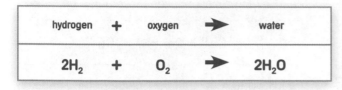

hydrogen	+	oxygen	➤	water
$2H_2$	+	O_2	➤	$2H_2O$

HT This **energy level diagram** shows the reaction between hydrogen and oxygen:

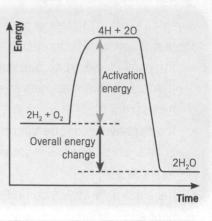

Fuel Cells

Hydrogen can be used as a fuel. When it reacts exothermically with oxygen in a **fuel cell** it makes an **electric current**. The energy from the reaction is used to create a **potential difference** (**pd**).

A fuel cell is very efficient at producing electrical energy. Fuel cells are used to provide electrical power in spacecraft. The water made is **pollution-free**, so it can be used as drinking water for the crew.

The car manufacturing industry is very interested in developing fuel cells as a possible pollution-free method of powering the electric car of the future. Currently, burning fossil fuels in cars produces carbon dioxide, which has been linked to climate change. Also, fossil fuels are non-renewable but there is a plentiful supply of hydrogen from the decomposing water.

HT In a fuel cell under **acidic** conditions:

1 Each hydrogen atom loses an electron at the **anode** (the positive electrode) to form a hydrogen ion. This is an example of **oxidation**:
$$H_2 - 2e^- ➤ 2H^+$$

2 The hydrogen ions then move through the electrolyte towards the **cathode** (the negative electrode) and the electrons travel around the circuit. Oxygen gets **reduced** as it gains electrons:
$$O_2 + 4H^+ + 4e^- ➤ 2H_2O$$

In a fuel cell under **alkaline** conditions:

1 At the **anode**, hydrogen is **oxidised**:
$$H_2 + 2OH^- - 2e^- ➤ 2H_2O$$

2 At the **cathode**, oxygen gets **reduced** as it gains electrons:
$$O_2 + 2H_2O + 4e^- ➤ 4OH^-$$

There are many advantages of using a fuel cell:
- They produce less pollution.
- They are very efficient.
- They transfer energy directly.
- They have few stages and are simple to construct.
- They are lightweight and compact.
- They have no moving parts.

HT There are some disadvantages of using fuel cells:
- They often contain poisonous catalysts which have to be carefully disposed of.
- To make the hydrogen fuel, energy is needed. This energy may come from the burning of fossil fuels.

In your exam you may be asked to explain the advantages of using a fuel cell compared to conventional methods for making electricity.

Rusting and Redox Reactions

Rust is a form of hydrated iron(III) oxide. It forms when iron or steel combines with both oxygen (in air) and water. This reaction is an example of a **redox reaction**. A redox reaction involves both **oxidation** and reduction. **Oxidation** is the gain of oxygen. **Reduction** is the removal of oxygen.

$$iron \; + \; oxygen \; + \; water \; \longrightarrow \; hydrated \; iron(III) \; oxide$$

(HT) Oxidation involves the loss of electrons. A chemical that removes electrons from another substance is called an **oxidising agent**.

Reduction involves the gain of electrons. A chemical that gives electrons to another substance is called a **reducing agent**.

Rusting is a redox reaction because:
- iron loses electrons (**oxidation**)
- oxygen gains electrons (**reduction**).

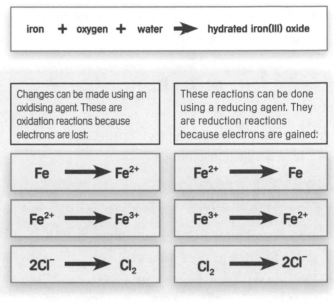

Changes can be made using an oxidising agent. These are oxidation reactions because electrons are lost:

$$Fe \longrightarrow Fe^{2+}$$
$$Fe^{2+} \longrightarrow Fe^{3+}$$
$$2Cl^- \longrightarrow Cl_2$$

These reactions can be done using a reducing agent. They are reduction reactions because electrons are gained:

$$Fe^{2+} \longrightarrow Fe$$
$$Fe^{3+} \longrightarrow Fe^{2+}$$
$$Cl_2 \longrightarrow 2Cl^-$$

Preventing Rusting

You can protect iron and steel from rusting by coating them in oil, grease or paint. This stops the water and air coming into contact with the metal. Other methods are as follows:
- **Galvanising** – coating the iron or steel with zinc. This layer stops water and oxygen reaching the iron. Also, zinc can act as sacrificial protection.
- **Alloying**.
- **Tin plating**.

- **Sacrificial protection** – placing a more reactive metal, for example, magnesium, in contact with the iron or steel.

(HT) Tin plating acts as a barrier between the iron and air and water. However, when the tin plating is scratched, the iron will corrode and lose electrons. In **sacrificial protection**, the more reactive metal, e.g. magnesium or zinc, will corrode, losing electrons and protecting the iron.

Displacement Reactions

A more reactive metal will displace a less reactive metal in a reaction. This type of reaction is known as a **displacement reaction**. Magnesium is more reactive than zinc, which is more reactive than iron, which is more reactive than tin.

You can write equations for displacement reactions, e.g.:

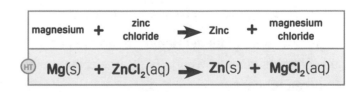

$$magnesium \; + \; zinc \; chloride \; \longrightarrow \; Zinc \; + \; magnesium \; chloride$$

(HT) $$Mg(s) \; + \; ZnCl_2(aq) \; \longrightarrow \; Zn(s) \; + \; MgCl_2(aq)$$

(HT) Displacement is a redox reaction. The metal ion is reduced by gaining electrons and the metal atom is oxidised by losing electrons.

C6 Alcohols

Ethanol and Alcohols

Alcohols are a family of organic compounds containing hydrogen, carbon and oxygen.

Ethanol (C_2H_5OH) is an alcohol. Its displayed formula is shown below:

$$H-\overset{\overset{\displaystyle H}{|}}{\underset{\underset{\displaystyle H}{|}}{C}}-\overset{\overset{\displaystyle H}{|}}{\underset{\underset{\displaystyle H}{|}}{C}}-O-H$$

Ethanol has many uses, for example, it can be used:
- to make alcoholic drinks
- to make solvents, such as methylated spirits
- as fuel for cars.

HT Alcohols have the general formula $C_nH_{(2n+1)}OH$. For example, pentanol is $C_5H_{11}OH$.

$$H-\overset{\overset{\displaystyle H}{|}}{\underset{\underset{\displaystyle H}{|}}{C}}-\overset{\overset{\displaystyle H}{|}}{\underset{\underset{\displaystyle H}{|}}{C}}-\overset{\overset{\displaystyle H}{|}}{\underset{\underset{\displaystyle H}{|}}{C}}-\overset{\overset{\displaystyle H}{|}}{\underset{\underset{\displaystyle H}{|}}{C}}-\overset{\overset{\displaystyle H}{|}}{\underset{\underset{\displaystyle H}{|}}{C}}-O-H$$

Making Ethanol by Fermentation

Ethanol can be made by **fermentation**. Yeast is used to ferment glucose solution.

glucose	➡	ethanol	+	carbon dioxide
$C_6H_{12}O_6$	➡	$2C_2H_5OH$	+	$2CO_2$

The apparatus used in fermentation prevents air (oxygen) from reaching the fermentation mixture.

The fermentation mixture has to be kept at between 25°C and 50°C for a few days. This is the **optimum temperature** for the enzymes in the yeast to change the glucose into ethanol. This is a renewable method.

Pure ethanol can be extracted from the fermentation mixture by **fractional distillation**.

HT The absence of air from fermentation prevents the formation of ethanoic acid by oxidation of the ethanol.

The temperature of the fermentation mixture has to be kept at between 25°C and 50°C (the optimum temperature) because:
- if it falls below the optimum temperature, the yeast becomes inactive
- if it rises above the optimum temperature, the enzymes in the yeast denature and stop working.

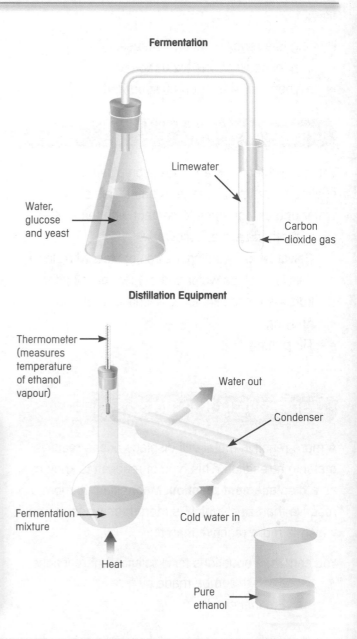

Fermentation

Limewater

Water, glucose and yeast

Carbon dioxide gas

Distillation Equipment

Thermometer (measures temperature of ethanol vapour)

Water out

Condenser

Fermentation mixture

Cold water in

Heat

Pure ethanol

Making Ethanol by Hydration

The chemical reaction to turn ethene into ethanol is **reversible**:

ethene $\xrightleftharpoons[\text{dehydration}]{\text{hydration}}$ ethanol

Ethene can be **hydrated** to make ethanol by passing it over a heated phosphoric acid catalyst with steam.

Ethanol made in this way is for industrial use only and is non-renewable.

ethene	+	water	phosphoric acid (catalyst)	ethanol
C_2H_4	+	H_2O	$\rightleftharpoons$	C_2H_5OH

Fermentation or Hydration?

The ethanol made by hydration is **non-renewable** because the ethene will have been made by cracking components of crude oil.

Ethanol made by fermentation is renewable and more sustainable.

HT The two methods of making ethanol have different advantages and disadvantages. These need to be considered before a company chooses which method to use.

It's quicker to produce ethanol by hydration than by fermentation, and ethene is available in large quantities in this country from cracking in oil refineries. This is a continuous process.

Making ethanol by hydration has a 100% atom economy. But fermentation has a higher percentage yield as the reaction is not reversible.

Fermentation is a slow batch process and the ethanol has to be purified by fractional distillation before use. This uses a lot of energy and is expensive.

Quick Test

1. What is galvanising?
2. What can ethanol be used for?
3. HT Ethanol can be made by fermenting. Is this production batch or continuous?
4. HT Ethanol can be made by hydration. Is this production batch or continuous?

C6 Depletion of the Ozone Layer

Depletion of the Ozone Layer

Ozone (O_3) is a type of oxygen found in a layer high up in the atmosphere (the stratosphere).

Chlorofluorocarbons (**CFCs**) are organic molecules that contain chlorine, fluorine and carbon. CFCs were used as refrigerants and in aerosols because:

- they have a low boiling point
- they are insoluble in water
- they are very unreactive (i.e. chemically inert).

CFCs damage the ozone layer. Hydrocarbons (alkanes) or hydrofluorocarbons (HFCs) are now used as safer alternatives to CFCs.

The depletion of ozone in the atmosphere allows increased levels of harmful ultraviolet (UV) light to reach the Earth, and this can cause:

- increased ageing of the skin and risk of sunburn
- skin cancers
- increased risk of cataracts.

When a CFC molecule is hit by ultraviolet light, a chlorine atom called a **radical** is produced.

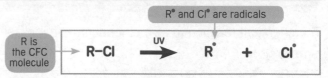

One chlorine radical can attack and destroy many ozone molecules. This leads to the loss of the ozone layer. Chlorine radicals and CFCs take a very long time to leave the stratosphere.

The international community banned the use of CFCs after they agreed with scientific researchers. The ban started in the developed countries, such as the UK. But, it was a few years before less-developed countries followed suit.

(HT) When CFCs were first made, they were seen to be very useful, particularly because of their inertness. But, the enthusiasm for their use was dampened when the link between ozone depletion and CFCs was found. Ozone depletion will continue for a few years yet as some countries are still using CFCs, and the inertness of CFCs causes them to stay in the environment for a long time.

(HT) CFCs' Effect on the Ozone Layer

Ozone filters out harmful ultraviolet light from reaching the surface of the Earth. When ozone absorbs ultraviolet light, the energy in the light causes the **covalent bond** in the ozone molecule to break. The ozone molecule is split into an oxygen molecule (O_2) and an oxygen atom (O). When this happens, the covalent bond can be broken evenly so that each atom retains one electron to form radicals, or unevenly to make **ions**.

When a CFC molecule is hit by ultraviolet light, the C—Cl bond breaks down. One electron from the covalent bond goes to the Cl to make a Cl radical and one electron goes to the main part of the molecule.

When a chlorine radical attacks an ozone molecule:

- the chlorine radical reacts with an ozone molecule to form a chlorine monoxide molecule and an oxygen molecule:

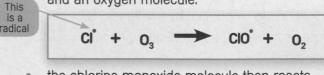

- the chlorine monoxide molecule then reacts with an oxygen atom to produce a chlorine radical and an oxygen molecule:

The chlorine radical is **regenerated** by this **chain reaction** and can go on to destroy many more ozone molecules.

    Ozone • Chlorofluorocarbons • Chain reaction

Hardness of Water

Water can be hard or soft.

- **Soft** water lathers well with soap.
- **Hard** water doesn't lather with soap.
- Both hard and soft water will lather with a soapless detergent.

Hardness in water is caused by calcium and magnesium ions from dissolved salts. There are two types of hardness in water:

- Permanent hardness.
- Temporary hardness.

Permanent hardness is caused by dissolved substances like calcium sulfate. It can't be destroyed by boiling.

Temporary hardness forms when rainwater comes into contact with rock that contains calcium carbonate, e.g. chalk, marble or limestone. It is caused by dissolved calcium hydrogencarbonate and it can be removed by boiling.

The hardness of the water depends on which rocks the water has flowed over. So the hardness of the water varies across the UK.

Forming and Removing Hardness

Rainwater contains dissolved carbon dioxide, which makes it slightly acidic. When rainwater reacts with rock it makes soluble calcium hydrogencarbonate. This causes temporary hardness:

| calcium carbonate | + | water | + | carbon dioxide | ➡ | calcium hydrogencarbonate |

When temporary hard water is boiled, the calcium hydrogencarbonate decomposes to form insoluble calcium carbonate, water and carbon dioxide, i.e. the hardness is removed from the water:

| calcium hydrogencarbonate | ➡ | calcium carbonate | + | water | + | carbon dioxide |

(HT) $Ca(HCO_3)_2(aq) \rightarrow CaCO_3(s) + H_2O(l) + CO_2(g)$

Limescale

Insoluble calcium carbonate will deposit as **limescale** on the heating element of an appliance, for example, in a kettle.

C6 Hardness of Water

Measuring Water Hardness

Measuring Water Hardness

Hardness in water can be measured by adding soap solution to the water until a permanent lather is produced after it is shaken for five seconds.

The number of drops of soap solution added can be counted, or a burette can be used to measure the volume of soap solution added.

In your exam, you will be expected to interpret data from an experiment to measure hardness in water.

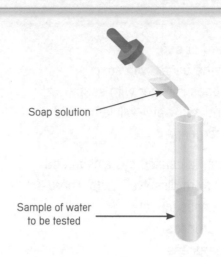

Soap solution

Sample of water to be tested

Removing Hardness from Water

You can remove all types of hardness from water by:

- adding washing soda (sodium carbonate crystals)
- passing the water through an **ion-exchange column**.

When hard water is passed through an ion-exchange column, the calcium and magnesium ions swap with the sodium ions, removing the hardness.

The column contains a resin with many sodium ions stuck to it. As the hard water passes over the resin, the calcium and magnesium ions attach to the resin and the sodium ions are released into the water.

(HT) When washing soda (sodium carbonate crystals) is used to soften water, the calcium and magnesium ions are removed from the water as they form precipitates of insoluble calcium carbonate and magnesium carbonate.

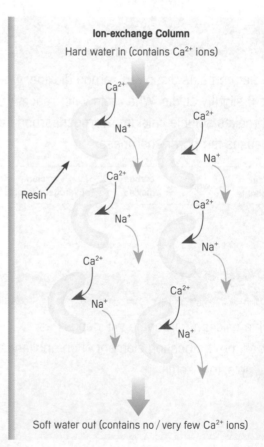

Ion-exchange Column

Hard water in (contains Ca^{2+} ions)

Ca^{2+}

Na^+

Ca^{2+}

Na^+

Ca^{2+}

Na^+

Ca^{2+}

Na^+

Resin

Ca^{2+}

Na^+

Ca^{2+}

Na^+

Soft water out (contains no / very few Ca^{2+} ions)

Quick Test

1. What elements are in chlorofluorocarbons (CFCs)?
2. Give an example of a radical that depletes the ozone layer.
3. Which dissolved ions cause the hardness in water?
4. Which two methods can be used to remove permanent hardness from water?

Ion-exchange column

Oils and fats

Oils and fats are **esters** that can be obtained from animals or vegetables.

At room temperature, oils are liquids and fats are solids.

Oils and fats can be:
- **saturated**, i.e. all the carbon-carbon bonds are single bonds (C–C)
- **unsaturated**, i.e. the molecule contains at least one carbon-carbon double bond (C=C).

You can shake an oil or fat with **bromine water** to test if it is saturated or unsaturated:
- If the fat is unsaturated, the bromine water will change from orange to colourless.
- If the fat is saturated, the bromine water will remain orange.

Oil and water are **immiscible**, i.e. they do not normally mix.

You can make an **emulsion** by shaking vegetable oil with water to break down the oil into small droplets that **disperse** (spread out) in the water.

For example:
- milk is an oil-in-water emulsion that is mostly water with tiny droplets of oil dispersed in it
- butter is a water-in-oil emulsion that is mostly oil with droplets of water dispersed in it.

HT Animal oils and fats are often saturated. Vegetable oils and fats are often unsaturated.

It is better for you to have more unsaturated than saturated oils and fats in your diet in order to reduce the build up of **cholesterol** in your body. This is healthier for the heart because cholesterol builds up in the blood vessels, causing the heart to do more work. It can eventually lead to heart disease.

When you are testing oils and fats using bromine water:
- saturated oils and fats do not react as they have no C=C, which is why the bromine water stays orange
- the double bonds in unsaturated oils and fats are able to undergo an addition reaction with the bromine water, so the bromine water turns colourless as a di-bromo compound is made.

Using Oils and Fats in Industry

Margarine can be made by reacting unsaturated vegetable oils with hydrogen, using a nickel catalyst. This makes a solid saturated fat, which can then be blended with other ingredients to make it taste and look like butter.

Soap can be made by reacting vegetable oil with hot sodium hydroxide. When this happens, the sodium hydroxide splits up the oil or fat molecules into **glycerol** and **soap** (a sodium salt of a long chain fatty acid). This is called **saponification.**

Oils and fats are very important raw materials for the chemical industry.

There is much interest in how vegetable oils can be converted into **biodiesel** to be used as a renewable replacement for diesel, which is obtained from **crude oil.**

HT This is the word equation for the process of saponification:

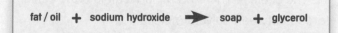

fat / oil + **sodium hydroxide** ➡ **soap** + **glycerol**

This reaction can be described as **hydrolysis** because it involves breaking down ester groups in the oil molecule using an alkali.

C6 Detergents

Washing Powder

The main components of a washing powder have specific jobs:

- **Active detergent** does the cleaning.
- **Bleach** removes coloured stains.
- **Water softener** softens hard water.
- **Optical brightener** makes whites appear brighter.
- **Enzymes** break up food and protein stains in low-temperature washes.

When clothes are washed, the:

- water is the solvent (the liquid that does the dissolving)
- washing powder is the **solute** (the solid that dissolves) because it's soluble (it dissolves) in water
- resulting mixture of solvent and solute is a **solution**.

Low-temperature washes are used because they:

- can be used to wash delicate fabrics (that would shrink in a hotter wash or have a dye that could run)
- don't denature enzymes in biological powders
- save energy.

Stain Removal

Different solvents dissolve different stains. The table shows which solvents can remove stains.

Although you don't need to learn this information, you might be asked to use similar information to choose which solvent to use to remove a stain. Some stains are **insoluble** (they will not dissolve) in water.

Dry cleaning solvents are used for these kinds of stain. The solvent is a liquid, but it doesn't contain water. This method of cleaning is used when a stain is insoluble in water.

Stain	Solvent
Ball-point pen	Methylated spirits (ethanol) then biological washing powder in water
Blood	Biological washing powder in water
Shoe polish	White spirit then biological washing powder in water
Coffee	Biological washing powder in water
Correcting fluid	White spirit

HT Dry-Cleaning Solvents

The molecules making up a stain are held together by **weak intermolecular forces** (forces between the molecules). There are also weak intermolecular forces between solvent molecules.

The stain will dissolve in a **dry-cleaning solvent** if the intermolecular forces holding it together are overcome. The new intermolecular forces between the stain molecules and the solvent molecules are stronger than the ones between the stain molecules.

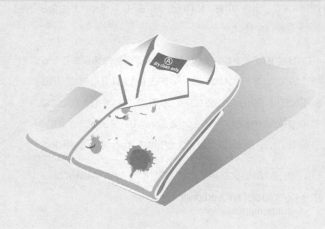

    Solvent • Soluble • Solution

Washing-Up Liquid

Washing-up liquid contains:

- **active detergent** – to do the cleaning
- **water** – to dissolve and dilute the detergent so that it's easy to pour
- **water softener** – to soften hard water
- **rinse agent** – to help the water drain off the crockery so it dries quickly
- **colour and fragrance** – to make the product more attractive to buy and use.

In your exam, you might be asked to interpret data from an experiment to suggest trends or which detergents washed the most plates.

HT You may be asked to interpret the data to suggest which detergents contain enzymes.

Detergent Molecules

A detergent molecule has a **hydrophilic** head and a **hydrophobic** tail:

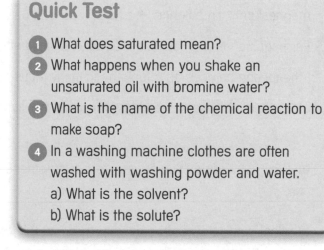

Detergent Molecule

Hydrophilic head

Detached sodium ion

Carbon

Sulfur

Hydrogen

Oxygen

Hydrocarbon tail

Hydrophobic tail

Quick Test

1. What does saturated mean?
2. What happens when you shake an unsaturated oil with bromine water?
3. What is the name of the chemical reaction to make soap?
4. In a washing machine clothes are often washed with washing powder and water.
   a) What is the solvent?
   b) What is the solute?

HT When detergent molecules dissolve in water:

1. The positively charged **sodium ion** comes away from the 'head' of the detergent molecule.
2. This leaves the molecule head negatively charged, so it's attracted to water molecules. It's known as **hydrophilic (water-loving)**.

3. The hydrocarbon tail is non-polar and so it isn't attracted to water molecules. It's known as **hydrophobic (water-hating)**.

This is how washing-up liquid detergents work:

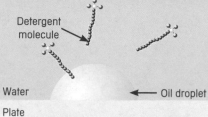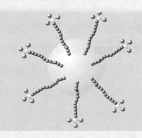

Detergent molecule

Water

Oil droplet

Plate

The hydrophobic end forms strong intermolecular forces with the oil molecules, causing it to stick to the oil droplet. The hydrophilic end forms strong intermolecular forces with the water.

As more and more detergent molecules are absorbed into the oil droplet, the oil is eventually lifted off the plate.

When it is totally surrounded, the oil droplet can be washed away, leaving the plate clean.

C6 Exam Practice Questions

1 Electrolysis breaks down an ionic compound into simpler substances.

a) During electrolysis, positive ions discharge at which electrode? **[1]**

b) How do you change an ionic solid into an electrolyte? **[2]**

c) What substance is made at the cathode during the electrolysis of copper(II) sulfate solution? **[1]**

2 A displacement reaction is when a more reactive metal takes the place of a less reactive metal in its compound.

Complete the word equation for this displacement reaction. **[2]**

magnesium + tin chloride → _____ + _____

3 Fermentation is a renewable method of making ethanol.

Ethanol can be used as a fuel and in many industrial processes as a solvent.

a) Explain why yeast is added during fermentation to produce ethanol. **[1]**

b) What type of reaction is used to convert ethene into ethanol? **[1]**

c) What catalyst would be used to carry out this reaction? **[1]**

4 **a)** The ozone layer is made of O_3 molecules and stops UV light hitting the surface of the Earth.

Explain why CFCs are harmful to the ozone layer. **[2]**

b) Which of the following are caused by being exposed to too much UV light?
Tick (✓) the correct answer(s). **[3]**

Skin cancer ⬭

Lung cancer ⬭

Sunburn ⬭

Cataracts ⬭

5 Hard water is made when dissolved magnesium and calcium ions are in water. Write a word equation to show how temporary hardness forms when limestone comes into contact with rainwater. **[2]**

6 Water can be described as hard when it contains certain dissolved metal ions. Explain what causes temporary and permanent hardness and how ion-exchange resin can be used to remove permanent hardness. **[6]**

✎ The quality of your written communication will be assessed in your answer to this question.

HT

7 Fat can undergo a chemical change with sodium hydroxide to form soap.

Write a word equation for the saponification of fat. **[1]**

8 Fuel cells can be used to power spacecraft.

a) Write a balanced symbol equation to show the reaction in a fuel cell. **[3]**

b) Explain why the overall reaction in a fuel cell is described as a redox reaction. **[2]**

9 Indicate whether the following reactions are **reduction** or **oxidation** by writing alongside each and explain your answers.

a) $Fe^{3+} \rightarrow Fe^{2+}$ _____ **[1]**

b) $Fe^{2+} \rightarrow Fe^{3+}$ _____ **[1]**

Answers

Fundamental Chemical Concepts

Quick Test Answers

Page 7
1. Electrons
2. In the nucleus.
3. A charged atom or group of atoms that has lost or gained electrons.
4. The different types of atom in a compound; The number of each type of atom; Where the bonds are in the compound.

C1 Carbon Chemistry

Quick Test Answers

Page 9
1. A resource that is being used up faster than it can be replaced.
2. Hydrogen and carbon.
3. Liquefied petroleum gas (LPG).
4. **a)** Breaking down a large hydrocarbon into smaller alkanes and alkenes. **b)** To increase the amount of the petrol fraction and to make alkenes that can be used for making polymers.

Page 11
1. Incomplete combustion (of a hydrocarbon).
2. Blue.
3. Oxygen.
4. $CH_4 + 2O_2 \rightarrow CO_2 + 2H_2O$
5. $2CH_4 + 3O_2 \rightarrow 2CO + 4H_2O$ **or** $CH_4 + O_2 \rightarrow C + 2H_2O$

Page 13
1. Nitrogen and oxygen.
2. Respiration and combustion.
3. Ammonia and carbon dioxide.
4. Photochemical smog and acid rain.

Page 17
1. A long-chain molecule made of small repeating units called monomers.
2. Poly(ethene).
3. **Any two from:** It is lightweight; It is waterproof; It is tough; It can easily be coloured.

Page 19
1. **Any one from:** Meat; Eggs; Fish.
2. **a)** Sodium hydrogencarbonate **b)** Thermal decomposition **c)** sodium hydrogencarbonate $\overset{heat}{\rightarrow}$ carbon dioxide + water + sodium carbonate **d)** carbon dioxide gas.
3. When protein molecules change shape as they are cooked.

Page 23
1. **Any suitable answer, e.g.** Lavender; Musk; Rose.
2. So that it does not wash off your skin easily.
3. A type of mixture.

4. Pigment, binding medium and a solvent.
5. To protect and decorate.
6. The solvent evaporates and the oil binding medium oxidises in the air.

Answers to Exam Practice Questions
1. **a)** The fractions can be separated and collected because hydrocarbons **[1]** boil at different temperatures **[1]**.
   b) Breaking up large hydrocarbon molecules into small hydrocarbon molecules **[1]**; To match supply and demand; To make more petrol; To make ethene **[Any one for 1]**.
   c) Small molecules have fewer forces of attraction between molecules than large molecules **[1]**; Less energy is needed to separate them **[1]**.
2. **a)** Nitrogen
   b) 21%
   c) **Any suitable answer e.g.** carbon dioxide.
3. **a)** All the fuel is burned in a blue flame (complete combustion); only some of the fuel's energy is released in a yellow flame (incomplete combustion).
   b) methane + oxygen $\rightarrow$ carbon dioxide + water
4. **a)** A small molecule that will join up to make a polymer.
   b) Catalyst; High pressure.
5. The material has pores **[1]** that are too small to allow water droplets through (from the outside) **[1]**, but are big enough to allow water vapour through (from the inside) **[1]**.
6. **This is a model answer, which demonstrates QWC and would therefore score the full 6 marks:** The emulsifier is made of a hydrophilic head and a hydrophobic tail. The hydrophilic head forms intermolecular forces of attraction with water molecules. The hydrophobic head forms intermolecular forces of attraction with the oil molecules. This allows the oil and water to remain mixed.

C2 Chemical Resources

Quick Test Answers

Page 28
1. Crust and top part of mantle.
2. They are less dense than the mantle.
3. Earthquakes and volcanoes.
4. Magma or lava that has solidified.
5. The slower the rock cools, the larger the crystals, or reverse argument.
6. Runny and has 'safe' eruptions.

Page 32
1. As ores.
2. **Any suitable answer, e.g.** Limestone; Marble; Granite.
3. Clay and limestone are heated together.

4. It is cheaper and can reduce the use of limited natural resources.
5. A mixture of a metal with another element.
6. It is transparent.
7. Water and oxygen (in the air).

Page 36
1. Nitrogen and hydrogen.
2. 450°C, 200 atmospheres and an iron catalyst.
3. A reaction that can go forwards or backwards under the same conditions.
4. A chemical with a pH of less than 7.
5. A chemical with a pH above 7, which can dissolve in water (a soluble base).
6. Acid + Metal carbonate $\rightarrow$ Salt + Water + Carbon dioxide

C2 Chemical Resources (Cont.)

Quick Test Answers
Page 38
1. A chemical that gives plants essential chemicals needed for growth.
2. Nitrogen, potassium and phosphorus.
3. Ammonia and sulfuric acid.
4. They use it to make protein.

Page 39
1. Sodium hydroxide, hydrogen and chlorine.
2. By reacting chlorine and sodium hydroxide together.
3. **Any suitable answer, e.g.** To sterilise water; To make household bleach; To make plastics (e.g. PVC); To make solvents.

Answers to Exam Practice Questions
1. The crust and the outer part of the mantle.
2. Slate – Natural
   Steel – Manufactured
   Cement – Manufactured
   Marble – Natural
   Brick – Manufactured
   [All correct for 3 marks, 4 correct for 2 marks, 2 or 3 correct for 1 mark]

3. a) It is heated with carbon.
   b) i) Copper(II) sulfate.
      ii) Pure copper.
      iii) Impure copper.
4. Air / oxygen; Water.
5. It saves resources **[1]** and reduces disposal problems **[1]**.
6. a) To increase the yield of their crop.
   b) Replaces essential elements used by the previous crop; Provides extra essential elements; More nitrogen gets into the plant protein, increasing its strength. **[Any two for 2]**
7. a) Air/oxygen; Water
   b) 72%
   c) 450°C
   d) 400 atmospheres
   e) Yield decreases
8. a) Ammonia solution; Ammonia hydroxide; Ammonia. **[Any one for 1]**
   b) $NH_3 + HNO_3 \rightarrow NH_4NO_3$ **[1 mark for each correct formula]**
9. Where a continental plate and an oceanic plate collide. The more dense oceanic plate is pushed under **[1]** the continental plate down into the mantle where it melts **[1]**. The result is a mountain range and possibly volcanoes **[1]**.

C3 Chemical Economics

Quick Test Answers
Page 45
1. A measure of the amount of product made in a specific time.
2. g/s or g/min.
3. Increase temperature, increase concentration (for reactants that are in solution) or increase pressure (for reactants that are gases) or add a catalyst.
4. The reactant particles have more energy (kinetic) and move around faster, making them more likely to collide and collisions are higher energy making them more likely to form a product.

Page 47
1. The mass of an atom compared with the mass of one twelfth of the mass of a carbon-12 atom.
2. The relative atomic masses of all the atoms in a compound, added together.

Page 50
1. $10 \div 44 \times 100 = 22.7\%$
2. Atom economy = M_r of desired products $\div$ sum of M_r of all products $\times$ 100
3. a) methane + oxygen → carbon dioxide + water
   b) Exothermic.

Page 53
1. **Any suitable answer, e.g.** Medicines; Pharmaceutical drugs.
2. Crushing, boiling, dissolving and chromatography.
3. Diamond, graphite and Buckminster fullerene.
4. C_{60}

Answers to Exam Practice Questions
1. a) i) $(2 \times 23) + 12 + (3 \times 16) = 106$ **[1 for calculation; 1 for correct answer]**
      ii) $(14 + [1 \times 4]) \times 2 + 32 + (4 \times 16) = 132$ **[2 for calculation; 1 for correct answer]**
   b) $9 \div 15 \times 100 = 60\%$ **[1 for calculation; 1 for correct answer]**

2. a) The materials for the new drug could be rare or may require expensive extraction from plants.
   b) You can make a product quickly on demand **[1]**; You can make a product on a small scale **[1]**; The equipment can be used to make a variety of products **[1]**.
3. a) They must collide **[1]** with sufficient energy **[1]**.
   b) There are more particles in the same volume, so they are more crowded **[1]**, and more particles collide more often **[1]**.
   c) The particles move faster **[1]**, causing them to collide more often and with more energy, resulting in successful collisions, and so an increase in rate of reaction. **[1]**
4. A powdered solid has a greater surface area in relation to its volume **[1]**. This increases the chance of collision **[1]** and speeds up the reaction. **[1]**
5. a) Time
   b) $40cm^3$
   c) **Accept any number between** 4.5 **and** 5 minutes.
6. a) Temperature; Concentration of solutions; Pressure of gases; Presence of a catalyst. **[Any two for 2]**
   b) Calculate the gradient at a given point of the line of best fit.
7. a) $4Fe + 3O_2 \rightarrow 2Fe_2O_3$ **[1 for correct reactants, 1 for correct products, 1 for correct balancing]**
   b) $(2 \times 56) + (3 \times 16) = 160$ **[1 for calculation, 1 for correct answer]**
   c) $4Fe + 3O_2 \rightarrow 2Fe_2O_3$ **[1]**
      $(4 \times 56) + 3 \times (16 \times 2) = 2 \times 160$ **[1 for correct calculation for reactions, 1 for correct calculation for products]**
      $320 = 320$ **[1]**

Answers

C4 The Periodic Table

Quick Test Answers

Page 58
1. Protons and neutrons.
2. A substance made of only one type of atom.
3. **a)** F **b)** 19 **c)** 9 **d)** 10

Page 63
1. Each sodium atom loses one electron from its outer shell to become a 1^+ ion.
2. The force of attraction between oppositely charged ions.
3. **a)** When molten or dissolved in water. **b)** MgO
4. A shared pair of electrons.

Page 65
1. All the elements have one electron in their outer shell.
2. Red

Page 69
1. All the elements have seven electrons in their outer shell.
2. Iron.
3. When a substance is broken down into two or more substances using heat.

Page 71
1. Run-off from fertilisers.
2. **a)** Barium chloride.
   b) sodium chloride + silver nitrate → silver chloride + sodium nitrate
3. **a)** Insoluble particles sink and can be removed.
   b) Chlorine.

Answers to Exam Practice Questions
1. **a)** A charged atom or group of atoms.
   b) Add 1 electron.
   c) High melting point; High boiling point; Conducts electricity in solution or when molten but not when solid.
2. **a)** Non-metals.
   b) There are no free-moving charged particles (electrons or ions) to carry the electricity.
   c) Group 4, period 3.
   d) They react with air **[1]** and water. **[1]**
   e) lithium + water → lithium hydroxide + hydrogen **[All correct for 2]**
3. **a)** Covalent bonding.
   b) 1 magnesium atom, 2 nitrogen atoms, 6 oxygen atoms.
   c) $2Na + 2H_2O \rightarrow 2NaOH + H_2$
4. **a)** Breaking up a substance **[1]** using heat. **[1]**
   b) copper carbonate → copper oxide + carbon dioxide **[All correct for 2]**
   c) Iron(III)
5. **a)** Chlorine.
   b) sodium iodide + chlorine → sodium chloride + iodine **[All correct for 2]**
   c) $2NaI + Cl_2 \rightarrow 2NaCl + I_2$ **[All correct for 2]**
   d) $Br_2 + 2e^- \rightarrow 2Br^-$ **[All correct for 2]**
6. **This is a model answer, which demonstrates QWC and therefore would score the full 6 marks:** The halogens have similar properties because when they react they all lose one electron to become a halide, negative ion. The halogens become less reactive as you go down the group. As you go down the group, the atoms become larger, with the outer shell electrons further away from the attractive positive force of the nucleus. This makes it more difficult for the atom to gain an additional electron.

C5 How Much (Quantitative Analysis)

Quick Test Answers

Page 75
1. A measurement of the amount of substance.
2. 44g/mol
3. **a)** 36g **b)** 18g/mol **c)** 4 moles

Page 76
1. CH
2. 53.5g/mol
3. CH_2

Page 78
1. $1.2 \, dm^3$
2. g/dm^3 or mol/dm^3
3. Add water to the solution.
4. If they are too dilute they may not work correctly and if they are too concentrated they may make you more ill.

Page 80
1. A chemical that changes colour depending on whether a chemical is acid or alkali.
2. The pH would start above pH 7 and slowly decrease. It will then decrease rapidly to below pH 7 and slowly decrease more until it reaches a steady level at a pH of less than 7.

Page 85
1. Measure the volume of gas made, or the mass lost.
2. ⇌
3. Temperature; Pressure; Concentration of reactants; Concentration of products.
4. $2SO_2 + O_2 \rightleftharpoons 2SO_3$

Page 87
1. An acid that completely ionises in water.
2. A precipitate is made when ions from two different solutions collide and react to make an insoluble product.
3. Barium nitrate.
4. Ethanoic acid does not completely ionise, whereas nitric acid does. This means that nitric acid releases more H^+ and so has a lower pH.

Answers to Exam Practice Questions
1. $0.15 \, dm^3$
2. The volume of acid or alkali added from a burette that just neutralises the test solution. (Final volume – Start volume on the burette)
3. It starts higher than 7 and falls lower **should be ticked**
4. $48 \, dm^3$ ($2 \times 12 = 24$) **[1 for calculation, 1 for correct answer]**
5. **a)** Yield increases
   b) Yield decreases
   c) 79%
6. **a)** Sulfur, air, water. **[All three named for 2]**
   b) The reaction would be too slow if it were cooler. It is a compromise between rate and yield.
7. A bromide.
8. **a)** $16 \div 40 = 0.4$ moles **[1 for calculation, 1 for correct answer]**
   b) $3 \times 16 = 48g$
9. $0.18 \, mol/dm^3$
10. $HCl \rightarrow H^+ + Cl^-$ **[All correct for 3]**
11. $KI + AgNO_3 \rightarrow KNO_3 + AgI$ **[All correct for 2]**

Answers

C6 Chemistry Out There

Quick Test Answers

Page 91
1. Negative electrode (cathode).
2. Size of current and the time it flows for.

Page 95
1. Coating iron or steel in zinc.
2. To make alcoholic drinks, solvents or as a fuel for cars.
3. Batch
4. Continuous

Page 98
1. Chlorine, fluorine and carbon.
2. A chlorine atom, Cl•
3. Magnesium and calcium ions.
4. Add washing soda (sodium carbonate crystals), or pass the water through an ion-exchange column.

Page 101
1. All the carbon-carbon bonds are single bonds.
2. The bromine water decolourises.
3. Saponification.
4. **a)** Water **b)** Washing powder

Answers to Exam Practice Questions

1. **a)** Cathode / negative electrode.
   b) Melt **[1]** or dissolve it. **[1]**
   c) Copper.
2. Tin; Magnesium chloride
3. **a)** This contains the enzymes to carry out the reaction.
   b) Hydration.
   c) Phosphoric acid.
4. **a)** CFCs break down and release chlorine atoms (radicals) **[1]** which destroy ozone molecules and make the ozone thinner. **[1]**
   b) Skin cancer; Sunburn; Cataracts **should be ticked**
5. calcium carbonate + water + carbon dioxide →
   calcium hydrogencarbonate **[All correct for 2]**
6. **This is a model answer, which demonstrates QWC and therefore would score the full 6 marks:** Temporary hardness is caused when rain water reacts with calcium carbonate to form dissolved sodium hydrogencarbonate. Dissolved magnesium and calcium ions cause permanent hardness. When hard water is passed through an ion-exchange resin, the calcium and magnesium ions are attracted to the resin and are swapped for sodium ions.
7. fat + sodium hydroxide → soap + glycerol
8. **a)** $2H_2 + O_2 \rightarrow 2H_2O$ **[All correct for 3]**
   b) Because the reactions involve both reduction **[1]** and oxidation. **[1]**
9. **a)** Reduction as electrons have been added.
   b) Oxidation as electrons have been removed.

Glossary of Key Words

Acid – a compound that has a pH value lower than 7.

Additives – chemicals added to food so that they look, taste or smell better or increase their shelf-life.

Alkali – a compound that has a pH value higher than 7 and can dissolve in water.

Alkane – a saturated (i.e. only single bonds) hydrocarbon.

Alkene – an unsaturated (i.e. at least one C=C) hydrocarbon.

Allotropes – different structural forms of the same element, e.g. diamond and graphite are both forms of carbon with different molecular structures.

Alloy – a mixture of two or more metals, or of a metal and a non-metal.

Anode – the positive electrode.

Atmosphere – the envelope of gas around the Earth.

Atom – the smallest part of an element that can enter into chemical reactions.

Atom economy – a measure of how many atoms from the reactants are in the desired product.

Atomic number – the number of protons in an atom; the number underneath the symbol in the periodic table.

Base – a compound that has a pH greater than 7 and that will neutralise an acid.

Batch process – a process where chemicals are added into a container, the reaction takes place, and the products are removed before a new reaction is started.

Calorimeter – a container used to hold liquids during a calorimetry experiment.

Calorimetry – an experiment used to measure the amount of energy released by a fuel or energy change in a reaction.

Catalyst – a substance that is used to speed up a chemical reaction without being chemically altered itself.

Cathode – the negative electrode.

Cement – a substance that sticks two other materials together; made by heating clay and limestone.

Chlorofluorocarbons (CFCs) – inert chemicals used as refrigerants and propellant gases; have been shown to damage the ozone layer.

Collision – when two or more particles hit each other.

Combustion – burning.

Complete combustion – burning in lots of oxygen.

Composite – a material made from two or more substances that can easily be seen, e.g. in plywood you can easily see the layers of the different wood.

Compound – a substance consisting of two or more elements chemically bonded.

Concentration – a measure of the amount of substance dissolved in a solution.

Concrete – a mixture of sand, gravel, water and cement.

Construction – building.

Contact Process – the process used to make sulfuric acid.

Continuous process – a process that doesn't stop; reactants are fed in at one end and products are removed at the other end at the same time.

Core – the centre of the Earth.

Corrosion – a reaction between metal and oxygen that turns it into a compound.

Covalent bond – a bond between two atoms in which one or more pairs of electrons are shared.

Cracking – breaking down (decomposition of) long chain hydrocarbons into smaller, more useful, short chain hydrocarbons.

Crust – the outer layer of the Earth.

Data – information collected from an experiment/investigation.

Decomposition – a chemical reaction where a compound breaks down into simpler substances.

Dilute – to reduce the concentration of a substance using water.

Electrode – the conducting rod or plate (usually metal or graphite) that allows electric current to enter and leave an electrolysis cell.

Electrolysis – the breaking down of a liquid or dissolved ionic substance using electricity.

Electrolyte – an aqueous or molten substance that contains free-moving ions and is therefore able to conduct electricity.

Electron – a negatively charged subatomic particle that orbits the nucleus of an atom.

Element – a substance that consists of only one type of atom.

Empirical formula – the simplest whole number ratio of each type of atom in a compound.

Emulsifiers – a chemical that can be used to make sure that oil and water remain mixed.

Emulsion – a mixture of oil (fat) and water.

Endothermic – a reaction that takes in energy.

Equilibrium – the state in which a chemical reaction proceeds at the same rate as its reverse reaction. (The quantities of reactants and products stay balanced.)

Ester – a family of chemicals that often smell nice.

Eutrophication – the excessive growth and decay of aquatic plants, e.g. algae, due to increased levels of nutrients in the water (often caused by fertilisers or untreated sewage), which results in oxygen levels dropping so that fish and other animal populations eventually die out.

Evaporation – a physical change where a liquid becomes a gas using the energy from other particles in that substance.

Exothermic – a reaction that releases energy.

Fermentation – the process by which yeast converts sugars to alcohol and carbon dioxide through anaerobic respiration.

Fertiliser – a chemical that helps plants grow and increases the yield of crops.

Fossil fuel – coal, oil or natural gas.

Fractional distillation – a method of separating a mixture of liquids each with a different boiling point.

Fuel cell – an electrochemical cell that converts chemical energy into electricity.

GDA (Guideline Daily Amount) – the recommended guidelines for daily amounts of nutrients.

Gore-Tex® – a brand name of a breathable layered fabric which includes a layer of Teflon™ polymer.

Group – a vertical column of elements in the periodic table.

Haber process – an industrial process where nitrogen and hydrogen are used to make ammonia.

Halide – a negative ion made from a Group 7 element that has gained one electron.

Halogens – elements in Group 7 of the periodic table.

Hardness – the inability of water to make a lather with soap due to dissolved ions in the water.

Hydrocarbon – a chemical containing only hydrogen and carbon.

Hydrophilic – water loving.

Hydrophobic – water hating.

Hydroxide – an OH^- ion.

Hypothesis – a scientific explanation that will be tested through experiments.

Incomplete combustion – burning in a limited supply of oxygen.

Indicator – a chemical that changes colour to show changes in pH.

Inert – does not undergo chemical reactions easily.

Insoluble – a substance that is unable to dissolve in a solvent.

Ion – a positively or negatively charged particle formed when an atom or group of atoms gains or loses one or more electron(s).

Ion-exchange column – a piece of equipment used to swap ions that cause hardness for sodium or hydrogen ions. This softens the water.

Ionic bond – the bond formed when electrons are transferred between a metal and a non-metal atom, creating charged ions that are then held together by forces of attraction.

Ionise – to make into ions.

Isotopes – atoms of the same element which contain different numbers of neutrons.

Lava – molten rock on the surface of the Earth.

Limescale – insoluble calcium carbonate deposits; found on the heating element of a kettle.

Limiting reactant – the reactant that gets used up first in a reaction.

Lithosphere – the crust and top part of the mantle of the Earth.

Magma – molten rock in the Earth.

Malleable – bends easily.

Mantle – the layer below the crust made of molten rock.

Mass number – the total number of protons and neutrons in an atom.

Model – a representation of a system or idea, used to describe or explain the system or idea.

Mole – a measurement of the amount of substance. Contains 6×10^{23} particles.

Monomer – a small unsaturated molecule that can be used to make a polymer.

Nanochemistry – the study of materials that have a very small size, in the order of 1–100nm; one nanometre is one billionth of a metre and can be written as 1nm or $1m \times 10^{-9}$.

Neutralisation – reaction between an acid and a base which forms a neutral solution.

Neutron – a sub-atomic particle found in the nucleus of atoms; it has no charge.

Non-renewable – a natural resource that is being used at a faster rate than it can be made.

Nucleus – the core of an atom, made up of protons and neutrons (except hydrogen, which contains a single proton).

Nylon – a brand name of a polymer.

Octet – eight electrons in the outer shell.

Ore – a raw material taken from the rocks of the Earth from which important chemicals can be extracted.

Oxidation – a reaction involving the gain of oxygen, the loss of hydrogen or the loss of electrons.

Ozone – an allotrope of oxygen made from O_3 molecules.

Period – a horizontal row of elements in the periodic table.

pH – a measure of the acidity of a solution.

Phosphorescent – a chemical that can store light and release it over a period of time.

Pigment – a coloured substance used in paints and dyes.

Pollutant – a chemical that can harm the environment and organisms.

Glossary of Key Words

Pollution – chemical contamination of the environment and / or organisms.

Polymer – a very long chain molecule with repeating units.

Polymerisation – the chemical reaction where monomers are used to make polymers.

Precipitate – an insoluble solid formed during a reaction involving ionic solutions.

Precipitation – the formation of an insoluble solid (a precipitate) when two solutions containing ions are mixed together.

Pressure – the amount of gas particles in a volume. It is like concentration for a gas.

Product – a substance made in a chemical reaction.

Proton – a positively charged sub-atomic particle found in the nucleus of an atom.

Reactant – a starting material in a reaction.

Recycle – to collect waste materials and make them into new products.

Redox reaction – a reaction that involves both reduction and oxidation.

Reduction – a reaction involving the loss of oxygen, the gain of hydrogen or the gain of electrons.

Relative atomic mass (A_r) – the mass of an atom compared to a twelfth of the mass of a carbon-12 atom.

Relative formula mass (M_r) – the sum of the atomic masses of all the atoms in a compound.

Reversible reaction – a reaction in which the products can react to reform the original reactants under the same conditions.

Rust – hydrated iron(III) oxide.

Rusting – the chemical reaction where iron reacts with oxygen from the air and water to make rust.

Salt – the product of a chemical reaction between a base and an acid.

Saponification – the process used to make soap by reacting vegetable oil with hot sodium hydroxide.

Seismic wave – the flow of energy going through the Earth after an earthquake.

Soluble – a property that means a substance can dissolve in a solvent.

Solute – the substance that gets dissolved.

Solution – the mixture formed when a solute dissolves in a solvent.

Solvent – a liquid that can dissolve another substance to produce a solution.

Stable – does not react.

Steel – an alloy mainly made from iron and a little carbon.

Strong acid – an acid that fully ionises when added to water.

Synthetic – man-made.

Tectonic plates – the large sections into which the Earth's crust is split.

Thermal decomposition – the use of heat to break down a substance into two or more substances.

Thermochromic – a chemical that changes colour as temperature changes.

Titration – a method used to find the concentration of an acid or an alkali.

Titre – the volume of acid needed to neutralise an alkali (or vice versa) in a titration.

Toxicity – how dangerous a chemical is to your health.

Universal indicator – a mixture of pH indicators, which produces a range of colours according to pH and can therefore be used to measure the pH of a solution.

Variable – something that changes during the course of an experiment/investigation.

Weak acid – an acid that partially ionises when added to water.

Yield – the amount of product obtained, e.g. from a crop or a chemical reaction.

HT **Chain reaction** – a reaction, e.g. nuclear fission, that is self-sustaining.

Distillation – a process used to separate liquids by evaporation followed by condensation to produce a pure liquid.

Electron configuration – the arrangement of electrons in the shell of an atom or ion.

Geology – the study of rocks and the structure of the Earth.

Saturated – has only single bonds.

Smart alloy – a mixture of two or more metals, or of a metal and a non-metal, that changes its properties as the environment changes.

Subduction – a plate boundary where one tectonic plate is forced below the other and the rock melts into the magma.

Unsaturated – has at least one double bond (C=C).

Notes

Notes

Notes

Key

| relative atomic mass |
| **atomic symbol** |
| name |
| atomic (proton) number |

1	
H	
hydrogen	
1	

Group 1	Group 2											Group 3	Group 4	Group 5	Group 6	Group 7	Group 0
1	2											3	4	5	6	7	0
																	4 **He** helium 2
7 **Li** lithium 3	9 **Be** beryllium 4											11 **B** boron 5	12 **C** carbon 6	14 **N** nitrogen 7	16 **O** oxygen 8	19 **F** fluorine 9	20 **Ne** neon 10
23 **Na** sodium 11	24 **Mg** magnesium 12											27 **Al** aluminium 13	28 **Si** silicon 14	31 **P** phosphorus 15	32 **S** sulfur 16	35.5 **Cl** chlorine 17	40 **Ar** argon 18
39 **K** potassium 19	40 **Ca** calcium 20	45 **Sc** scandium 21	48 **Ti** titanium 22	51 **V** vanadium 23	52 **Cr** chromium 24	55 **Mn** manganese 25	56 **Fe** iron 26	59 **Co** cobalt 27	59 **Ni** nickel 28	63.5 **Cu** copper 29	65 **Zn** zinc 30	70 **Ga** gallium 31	73 **Ge** germanium 32	75 **As** arsenic 33	79 **Se** selenium 34	80 **Br** bromine 35	84 **Kr** krypton 36
85 **Rb** rubidium 37	88 **Sr** strontium 38	89 **Y** yttrium 39	91 **Zr** zirconium 40	93 **Nb** niobium 41	96 **Mo** molybdenum 42	[98] **Tc** technetium 43	101 **Ru** ruthenium 44	103 **Rh** rhodium 45	106 **Pd** palladium 46	108 **Ag** silver 47	112 **Cd** cadmium 48	115 **In** indium 49	119 **Sn** tin 50	122 **Sb** antimony 51	128 **Te** tellurium 52	127 **I** iodine 53	131 **Xe** xenon 54
133 **Cs** caesium 55	137 **Ba** barium 56	139 **La*** lanthanum 57	178 **Hf** hafnium 72	181 **Ta** tantalum 73	184 **W** tungsten 74	186 **Re** rhenium 75	190 **Os** osmium 76	192 **Ir** iridium 77	195 **Pt** platinum 78	197 **Au** gold 79	201 **Hg** mercury 80	204 **Tl** thallium 81	207 **Pb** lead 82	209 **Bi** bismuth 83	[209] **Po** polonium 84	[210] **At** astatine 85	[222] **Rn** radon 86
[223] **Fr** francium 87	[226] **Ra** radium 88	[227] **Ac*** actinium 89	[261] **Rf** rutherfordium 104	[262] **Db** dubnium 105	[266] **Sg** seaborgium 106	[264] **Bh** bohrium 107	[277] **Hs** hassium 108	[268] **Mt** meitnerium 109	[271] **Ds** darmstadtium 110	[272] **Rg** roentgenium 111							

Elements with atomic numbers 112–116 have been reported but not fully authenticated

*The lanthanoids (atomic numbers 58–71) and the actinoids (atomic numbers 90–103) have been omitted.

Index